T0136785

FLORA ZAMBESIACA

Flora terrarum Zambesii aquis conjunctarum

VOLUME EIGHT: PART THREE

FLORA ZAMBESIACA

MOZAMBIQUE

MALAWI, ZAMBIA, ZIMBABWE

BOTSWANA

VOLUME EIGHT: PART THREE

Edited by

E. LAUNERT

on behalf of the Editorial Board:

E. A. BELL
Royal Botanic Gardens, Kew

E. LAUNERT
British Museum (Natural History)

M. L. GONÇALVES
*Centro de Botânica, Instituto de Investigação
Científica Tropical, Lisboa*

Published by the Managing Committee on behalf of
the contributors to Flora Zambesiaca
1988

© Flora Zambesiaca Managing Committee, 1988

Typeset by TND Serif Ltd Hadleigh, Suffolk
Printed in Great Britain by Halesworth Press Ltd

ISBN 0 9507682 5 1

CONTENTS

LIST OF FAMILIES INCLUDED IN
VOLUME VIII, PART 3

LIST OF NEW TAXA PUBLISHED
IN THIS WORK

Sesamum calycinum Subsp. *baumii* (Stapf) Seidenst. Stat. nov.
Sesamum calycinum Subsp. *pseudoangolense* Seidenst. Subsp. nov.

122. LENTIBULARIACEAE

By P. Taylor

Herbaceous terrestrial, epiphytic or aquatic plants all with specialized organs for the capture of small organisms. Roots frequently absent. Leaves rosulate or scattered on stolons, entire or divided, sometimes polymorphic. Inflorescence terminal or lateral, peduncled, racemose, bracteate, or solitary, scapose; lowest bracts (scales) usually barren; bracteoles 2 or absent or more or less connate with the bract, usually at the base of the pedicels. Flowers hermaphrodite, zygomorphic. Calyx deeply 2−4 or 5-partite, almost regular or more or less bilobed or the sepals free to the base, persistent and often accrescent. Corolla gamopetalous, bilabiate usually spurred, rarely saccate, usually violet or yellow; tube very short; superior lip interior, more or less entire or 2 or rarely more lobed; inferior lip entire or 2−5-lobed, usually with a raised more or less gibbous palate. Stamens 2, anticous, inserted at the base of the corolla; filaments usually short, usually curved, rarely longer and geniculate; anthers 2-thecous; thecae diverging, usually more or less confluent, dehiscing by a common slit. Ovary superior, unilocular; carpels 2, median; style simple, usually short or very short, rarely longer and geniculate; stigma more or less bilabiate, the superior lip usually smaller than the inferior or more or less obsolete; placenta free, basal, usually ovoid or globose; ovules usually numerous, sessile, rarely fewer or 2, anatropous. Fruit a unilocular capsule, 1-many-seeded, dehiscing by longitudinal slits or by pores or circumscissile or rarely indehiscent. Seeds small or very small, very variously shaped; testa thin or spongy or corky, rarely mucilaginous; embryo undifferentiated.

A family of 3 genera and about 280 species, mainly in the tropics.

Calyx lobes 2 (or 4 in some Australian species) -   -   -   - ⟩ 1. Utricularia
Calyx lobes 5 -   -   -   -   -   -   -   -   -   -   - 2. Genlisea

1. UTRICULARIA L.

Utricularia L., Sp. Pl: 18 (1753) & Gen. Pl., ed. 5: 11 (1754).—P. Taylor in Kew Bull. 18: 1 (1964); in Fl. Afr. Centr., Lentibulariaceae: 2 (1972); in F.T.E.A., Lentibulariaceae: 1 (1973).

Herbs, annual or perennial, terrestrial, epiphytic or aquatic. Vegetative parts not clearly differentiated but consisting of stems modified to function as roots, stems, leaves and specialized organs (traps) for the capture of small organisms. Root-like organs (rhizoids) usually descending from the base of the inflorescence, usually filiform. Stem-like organs (stolons) arising with the rhizoids at the inflorescence base; in the terrestrial and epiphytic species usually short and delicate but sometimes developing into fleshy tubers; in the aquatic species usually more robust and longer. Foliar organs (leaves) either rosulate at the inflorescence base or alternate, opposite or verticillate on the stolons; in the terrestrial and epiphytic species mostly entire, erect or thalloid, capillary, linear, circular or peltate; in the aquatic species usually more or less dichotomously divided into capillary segments. Traps rosulate, or lateral on the rhizoids, stolons or leaves or rarely terminal on the leaves, hollow, globose or ovoid, usually stalked, with a mouth which may be basal (adjacent to the stalk), lateral or terminal (opposite to the stalk); mouth usually provided externally with very diverse appendages. Inflorescence racemose, (but sometimes with a single flower) bracteate; peduncle usually simple, sometimes branched, usually filiform, erect or twining, usually glabrous, sometimes papillose, glandular or hairy, often provided (especially in the terrestrial species) with sterile bracts (scales); raceme usually elongated, rarely short and subcapitate; pedicels usually short, terete, flattened or more or less winged, often deflexed or decurved in fruit; bracts persistent, basifixed or produced below the point of insertion or peltate; bracteoles 2 or absent or sometimes more or less fused with the bract, inserted with the bract at the base of the pedicel or rarely with the calyx lobes at the apex of the pedicel. Calyx lobes 2, or rarely 4 in 2 decussate pairs, more or less equal or sometimes very unequal, usually free, sometimes more or less united at the base, persistent and usually accrescent, sometimes very markedly so; upper lobe usually entire,inferior lobe usually emarginate or bidentate, rarely both lobes dentate or fimbriate. Corolla bilabiate, glabrous, glandular or pubescent; throat closed or sometimes open; superior lip usually more or less erect, limb entire, emarginate or bilobed; inferior lip usually larger, spurred or rarely saccate at the base, palate usually raised and gibbous, limb spreading or deflexed, entire, emarginate or more or less deeply 2—5-lobed; spur more or less parallel to the inferior lip or divergent at an acute or obtuse angle or rarely in the same plane. Stamens 2, inserted at the base of the corolla; filaments straight or curved, usually twisted, sometimes winged; anthers dorsifixed, more or less ellipsoid, the thecae usually more or less confluent. Ovary globose or ovoid, unilocular; style usually short, often indistinct, persistent; stigma bilabiate, inferior lip usually much larger than the upper which may be obscure or obsolete; ovules 2-many, sessile on a more or less fleshy free basal placenta, anatropous. Fruit a capsule, more or less globose or ovoid, dehiscing by longitudinal slits or by pores or circumscissile or indehiscent. Seeds 1-many, usually small or very small, globose, ovoid, truncate conical, narrowly cylindrical, fusiform, lenticular or prismatic, smooth, verrucose, reticulate, glochidiate, papillose or comose or variously winged.

A large genus of c. 215 species distributed throughout the tropics with a few in temperate or even arctic regions. 38 species are recorded from tropical and S. Africa of which 28 occur in the Flora Zambesiaca area.

1. Margin of calyx lobes dentate or fimbriate   -   -   -   -   -   -   2
 — Margin of calyx lobes entire (apex may be emarginate or bi- or tridentate) -   3
2. Bracts and bracteoles fimbriate; bracteoles inserted at the apex of the pedicel; corolla yellow -   -   -   -   -   -   -   -   -   -   11. simulans
 — Bracts and bracteoles entire; bracteoles inserted at the base of the pedicel; corolla violet -   -   -   -   -   -   -   -   -   -   6. odontosepala
3. Calyx, bracts, pedicels and spur of the corolla covered with gland-tipped hairs
    -   -   -   -   -   -   -   -   -   -   -   -   -   10. podadena
 — Calyx, bracts and pedicels without gland-tipped hairs ..   -   -   -   4
4. Bracts produced below the point of attachment   -   -   -   -   -   5
 — Bracts not produced below the point of attachmnent   -   -   -   -   9
5. Bracteoles absent; fertile bracts circular or elliptic, amplexicaul; corolla yellow
    -   -   -   -   -   -   -   -   -   -   -   -   20. subulata
 — Bracteoles present; fertile bracts lanceolate or ovate   -   -   -   -   6
6. Bracteoles similar to the bracts, produced below the point of attachment; corolla white or violet -   -   -   -   -   -   -   -   -   -   -   7
 — Bracteoles differing from the bracts, not produced below the point of attachment; corolla yellow   -   -   -   -   -   -   -   -   -   1. bracteata
7. Leaves always present and conspicuous at anthesis, reniform or circular, petiolate; inferior lip of corolla 5-lobed; seeds glochidiate   -   -   -   18. striatula
 — Leaves rarely present or conspicuous at anthesis, peltate or linear; seeds not glochidiate -   -   -   -   -   -   -   -   -   -   -   8
8. Calyx lobes more or less equal, narrowly ovate; peduncle erect; leaves, when present, peltate -   -   -   -   -   -   -   -   -   -   9. pubescens
 — Calyx lobes unequal, the upper broadly obovoid, the lower oblong-ovate, much smaller; peduncle twining; leaves, when present, linear -   -   -   -   19. appendiculata
9. Bracteoles present; affixed terrestrial or aquatic plants -   -   -   -   10
 — Bracteoles absent; more or less freely suspended aquatic plants   -   -   21
10. Bracts more than 4 times as wide as the bracteoles and usually more or less concealing them; pedicels usually flattened or winged and usually as long as or longer than the fruiting calyx   -   -   -   -   -   -   -   -   -   -   -   -   11
 — Bracts not more than twice as wide as the bracteoles and similar in shape; pedicels terete, usually shorter than the fruiting calyx -   -   -   -   -   -   16
11. Fruiting pedicels deflexed; corolla mauve or violet, about 4 mm. long
    -   -   -   -   -   -   -   -   -   -   -   -   17. baoulensis
 — Fruiting pedicels erect or spreading .   -   -   -   -   -   -   -   12
12. Seeds globose, reticulate, the cells of the testa more or less isodiametric; capsule wall membranous throughout; corolla violet   -   -   -   -   -   -   13
 — Seeds ovoid or oblong; the cells of the testa distinctly elongate; capsule wall with a narrow thickened area on either side of the line of dehiscence; corolla yellow
    -   -   -   -   -   -   -   -   -   -   -   -   -   -   14

13. Superior lip of the corolla not or only slightly exceeding the upper calyx lobe; corolla up to 10 mm. long . . . . 16. *tortilis*
  − Superior lip of the corolla longer and 2−3 times as wide as the upper calyx lobe; corolla 12−30 mm. long . . . . 15. *spiralis*
14. Seeds smooth with a more or less prominent hilum; leaves linear with a single nerve . . . . 13. *scandens*
  − Seeds verrucose; leaves 3-or more-nerved 15
15. Corolla 15−20 mm. long, 4-carinate at the base of the inferior lip; leaves often not present at anthesis, scattered on the stolons, not forming a rosette; seeds 0.6− 0.7 mm. long; traps with a ventral appendage . . . . 14. *prehensilis*
  − Corolla less than 10 mm. long, not 4-carinate at the base of the inferior lip; leaves usually present and forming a rosette; seeds 0.4−0.5 mm. long; traps without a ventral appendage . . . . 12. *andongensis*
16. Inferior lip of corolla more or less deeply 5-lobed, superior lip tapering gradually from the base to the more or less bifid apex; peduncle papillose or hispid at the base . . . . 4. *pentadactyla*
  − Inferior lip of corolla not deeply 5-lobed, superior lip more or less constricted between base and apex 17
17. Superior lip of corolla with the lower part tapering gradually into the narrowly oblong or obovoid superior part, which is always broadest near the apex and usually much narrower than the calyx; palate of inferior lip usually with transversely wrinkled marg - 18
  − Superior lip of corolla with lower part ovate, suddenly constricted at the junction with the superior part which is always broadest below the middle and usually broader than the calyx; palate of the inferior lip without transversely wrinkled margin 19
18. Inferior lip of corolla about half as long as the spur; corolla 3−9 mm. long; calyx not plicate; base of peduncle usually papillose . . . . . 2. *arenaria*
  − Inferior lip of corolla about two thirds to equal or slightly longer than the spur; corolla 6−15 mm. long; calyx more or less plicate; base of peduncle always glabrous . . . . 3. *livida*
19. Calyx lobes with strongly incurved margins, densely papillose; corolla more or less persistent - 20
  − Calyx lobes without strongly incurved margins, smooth or nearly so; corolla not persistent - . . . . 7. *microcalyx*
20. Spur 2−3 times as long as the 3-lobed inferior lip of the corolla, straight; inflorescence axis straight; corolla yellow, 4−5 mm. long - . . . . . 8. *firmula*
  − Spur about 1.5 times as long as the inferior lip of the corolla, curved; inflorescence axis usually flexuous; corolla violet or rarely abnormally yellow, 6−15 mm. long . . . . 5. *welwitschii*
21. Inflorescence with a whorl of inflated floats 22
  − Inflorescence without a whorl of inflated floats 24
22. Stolons and traps villous; calyx minute, much shorter than the fruit; spur of the corolla inflated, much longer than the inferior lip; seeds lenticicular with a narrow irregular wing . . . . 25. *benjaminiana*
  − Stolons and traps glabrous 23
23. Lobes of fruiting calyx erect, concealing the capsule; seeds about as wide as high; floats narrowly cylindrical-fusiform, usually about 6−8 times as long as broad; corolla usually white or mauve - . . . . 22. *inflexa*
  − Lobes of fruiting calyx reflexed, exposing the capsule; seeds 2 or more times as wide as high; floats narrowly ellipsoid, usually about 2−4 times as long as broad; corolla usually yellow - . . . . 23. *stellaris*
24. Spur very short, saccate; corolla white or cream, less than 4 mm. long; inflorescence less than 1.5 cm. high . . . . 28. *cymbantha*
  − Spur well developed; corolla yellow, at least 5 mm. long; inflorescence more than 3 cm. high - 25
25. Traps produced in the angles of the leaf bifurcations; corolla usually hairy on the outside . . . . 24. *reflexa*
  − Traps produced laterally on the leaf segments; corolla not hairy on the outside 26
26. Leaves with few bifurcations; inflorescence 1−4-flowered . . 27. *gibba*
  − Leaves with very numerous bifurcations, inflorescence 5-many flowered - 27
27. Stolons terete with internodes 0.5−1 cm. long; leaves up to 3 cm. long, all similar and bearing traps . . . . 21. *australis*
  − Stolons flattened, with internodes 3−10 cm. long; leaves up to 15 cm. long, dimorphic, some not bearing traps . . . . 26. *foliosa*

1. **Utricularia bracteata** Good in Journ. Bot. 62: 161 (1924).−P. Taylor in Kew Bull. 18: 99 (1964); Fl. Afr. Centr., Lentibulariaceae: 28 (1972). Type from Angola.

Terrestrial herb. Rhizoids and stolons numerous from the base of the peduncle. Leaves numerous, scattered on the stolons, linear-spathulate, up to 20 × 1 mm., 1-nerved. Traps dimorphic, shortly stalked, ovoid, those on the stolons c. 1.5 mm. long, those on the stolon branches c. 0.5 mm. long; mouth terminal, fringed with radiating rows of glandular hairs, the larger traps with a long carinate dorsal beak which is reduced, in the smaller traps, to a short deltoid tooth. Inflorescence erect, up to 30 cm. high; peduncle filiform, glabrous; flowers 2−5, congested; scales few, similar to the bracts; bracts medifixed, lanceolate, up to 6 mm. long; bracteoles basifixed, oblong, 2−3 mm. long; pedicels filiform, erect, 1−3 mm. long. Calyx lobes subequal c. 5 mm. long; superior lobe circular with apex rounded; inferior lobe broadly obovate with apex emarginate. Corolla yellow, 10−13 mm. long; superior lip oblong with apex truncate or emarginate, about twice as long as the superior calyx lobe; inferior lip circular; palate slightly raised; spur conical, slightly longer than the inferior lip. Filaments linear; anther thecae subdistinct. Ovary ovoid, style indistinct, inferior lip of stigma circular, the superior much smaller, deltoid. Capsule globose, c. 3 mm. long, dehiscing by a single longitudinal ventral slit. Seeds numerous, ovoid, c. 0.35 mm. long, verrucose; testa cells distinct, elongate with sinuate margins.

Zambia. W: Chizela, fl. 11.vi.1953, *Fanshawe 78* (BR; EA; K; SRGH).
Also in Central African Republic and Angola. Wet peaty grassland.

2. **Utricularia arenaria** A.DC. in DC., Prodr. 8: 20 (1844).−Oliver in Journ. Linn. Soc., Bot. 9: 153 (1865).−P. Taylor in F.W.T.A., ed. 2, 2: 378 (1963); in Kew Bull. 18: 107 (1964); in Fl. Afr. Centr., Lentibulariaceae: 22 (1972); in F.T.E.A., Lentibulariaceae: 11 (1973). Type from Senegal.
Utricularia tribracteata Hochst. ex A. Rich., Tent. Fl. Abyss. 2: 18 (1851).−Kam. in Engl., Bot. Jahrb. 33: 99 (1902), partim, quoad spec. Schimper 1943.−Stapf in Hook., Ic. Pl. 28, sub. t. 2795B (1905), partim, quad spec. Schimper 1943, excl. fig. & syn.−Dyer, F.T.A. 4, 2: 475 (1906), partim, quad spec. Schimper 1943 & excl. syn.− Hutch. & Dalz., F.W.T.A. 2: 234 (1931).−F.W. Andr., Fl. Pl. Sudan 3: 153 (1956), partim, excl. spec. "Imatong Mts.". Type from Ethiopia.
Utricularia exilis Oliver in Journ. Linn. Soc., Bot. 9: 154 (1865).−Hiern, Cat. Afr. Pl. Welw. 1: 788 (1900).−Kam. in Engl., Bot. Jahrb. 33: 97 (1902), partim excl. spec. Ecklon & Schlechter, in Warb., Kunene-Samb.-Exped. Baum: 372 (1903).−Stapf in Hook., Ic. Pl. 28, t. 2797B (1905).−Dyer, F.T.A. 4, 2: 477 (1906).−Pellegrin in Bull. Soc. Bot. Fr. 61: 16 (1914).−Dinter in Fedde Repert. 24: 367 (1928).−Peter, Wasserpfl. Deutsch-Ostafr.: 127 (1928).−Pellegrin, Fl. Mayombe 2: 43 (1928).−Good in Journ. Bot. 68, Suppl. Gamopet.: 123 (1930).−F.W. Andr., Fl. Pl. Sudan 3: 153 (1956). Syntypes from Angola.
Utricularia exilis var. *bryoides* Welw. ex Hiern, Cat. Afr. Pl. Welw. 1: 789 (1900). Type from Angola.
Utricularia exilis var. *nematoscapa* Welw. ex Hiern, loc. cit.: 789 (1900). Type from Angola.
Utricularia exilis var. *ecklonii* (Spreng.)Kam. in Engl., Bot. Jahrb. 33: 98 (1902), partim, quoad spec. Kirk 2.
Utricularia exilis var. *hirsuta* Kam., loc. cit.: 98 (1902). Type from Namibia.
Utricularia exilis var. *arenaria* (A.DC.) Kam., loc. cit.: 98 (1902), partim, quoad spec. Perrottet.
Utricularia kirkii Stapf in Dyer, F.C. 4, 2: 428 (1904).−Hook., Ic. Pl. 28, t. 2795 (1905), partim, excl. spec. Kirk "Batoka Country".−Dyer, F.T.A. 4, 2: 476 (1906), partim, excl. spec. Kirk "Batoka Country".−Burtt Davy & Pott-Leendertz, Check-list Fl. Pl. Transv.: 164 (1912).−Dinter in Fedde, Repert. 24: 367 (1928).−Peter, Wasserpfl. Deutsch-Ostafr.: 127 (1928). Syntypes from Transvaal.
Utricularia monophylla Dinter in Fedde, Repert. 24: 367 (1928) e descr. Type from Namibia.
Utricularia parkeri sensu Perrier in Humbert, Fl. Madag. Lentibulariaceae: 16 (1955), partim, quoad spec. Perrier 8588, 16258, Viguier & Humbert 1807 non Baker.
Utricularia ecklonii sensu Perrier in Mém. Inst. Sci. Madag., Sér. B, 5: 198 (1955).− Humbert, Fl. Madag. Lentibulariaceae: 17 (1955), partim, quoad spec. Perrier 8384, 8587, 8598, 14784, 14789, non Spreng.

Terrestrial herb. Rhizoids and stolons capillary, numerous from the base of the peduncle. Leaves usually numerous at anthesis, scattered on the stolons, linear-oblanceolate to obovate-spathulate, 2−15 mm. long, 0.3−2 mm. wide, 1-nerved. Traps numerous, ovoid, stalked, 0.6−1.0 mm. long; mouth terminal provided with dorsal and ventral radiating rows of glandular hairs. Inflorescence erect, 2−16 cm. high; flowers 1−5(8), distant; scape filiform, smooth above, usually papillose at the base; scales few, similar to the bracts; bracts basifixed, ovate-lanceolate, acute, more or less 1 mm. long; bracteoles similar but narrower; pedicels 0.5−1 mm. long. Calyx lobes subequal, the upper slightly larger, broadly ovate, acute, lower ovate-oblong, rounded or truncate. Corolla white or lilac with a yellow spot on the palate, 3−7 mm. long; superior lip narrowly oblong, more or less 1.5 times as long as the upper calyx lobe, with apex rounded, truncate or emarginate; inferior lip circular; palate raised, double-crested, the crest smooth or transversely tuberculate; spur conical-subulate, more or less twice as long as the inferior lip. Filaments filiform; anther-thecae subdistinct. Ovary ovoid; style very short; inferior lip of stigma semi-circular, the superior deltoid, much smaller. Capsule globose, 1.5−2.5 mm. in diam., dehisceing by longitidinal dorsal and ventral slits with thickened margins. Seeds numerous, more or less 0.2 mm. long, truncate-conical, angular; testa smooth, cells indistinct, oblong.

Caprivi Strip: Grootfontein N., fl. 11.iii.1958, *Merxmüller* 2102 (BR). **Botswana.** N: Moremi Game Reserve, Xakanax Lediba, fl. 28.vi.1975, *Smith* 1409 (K; SRGH). **Zambia.** N: Mbala Distr., Kambole Escarpment, 1500 m., fl. 29.i.1964, *Richards* 18884 (K). W: Kasempa Distr., 7 km. E. of Chizela, fl. 27.iii.1961, *Drummond & Rutherford-Smith* (K; SRGH). C: c. 10 km. E. of Lusaka, c. 1250 m. fl. 24.v.1955, *King* 4 (K). E: 3 km. W. of Kachalola on Great E. Rd., c. 500 m. fl. 17.iii.1959, *Robson* 1733 (B; BR; EA; K; LISC; LM; PRE; SRGH). S: Gwembe Distr., c. 0.8 km. from Chirundu Bridge, near Lusaka Rd., fl. 5.ii.1958, *Drummond* 5479 (BR; K; SRGH). **Zimbabwe.** W: Matobo, Farm Besna Kobila, c. 1550 m. fl. ii.1954, *Miller* 2140 (K; SRGH). C: Harare, Cleveland Dam, fl. 19.v.1953, *Wild* 4115 (K; MO; SRGH). E: Mutare Distr., Rowa area, Zimunya Reserve, 1233 m., 16.vii.1961, *Chase* 7509 (K; SRGH). S: Bikita Distr., Turgwe-Dafana confluence, 1050 m., fl. 5.v.1969, *Biegel* 3019 (K). **Malawi.** C: Kasungu Distr., c. 10 km. S. of Kasungu on M1, 1250 m., fl. 7.iv.1978, *Pawek* 14337 (K; MAL; MO; SRGH; UC). **Mozambique.** Z: Quelimane Distr., fl. 3.vii.1949, *Faulkner* K444 (K). MS: Biera Distr., Cheringoma, lower Chinziua Saw Mill c. 50 m., fl. 13.vii.1972, *Ward* 7890 (EA; K). GI: Gaza, Bilene, edge of lagoon, fl. 5.xii.1979, *Schafer* 7114 (K; LIJSC).
Widespread in tropical Africa from Senegal to Sudan and Ethiopia to Angola, also in Namibia, Transvaal, Natal, Madagascar and India. Marshes and swampy grassland and shallow sandy or peaty soil overlying rocks; sea level to 2400 m.

3. **Utricularia livida** E. Mey., Comm. Pl. Afr. Austr.: 281 (1837).—Oliver in Journ. Linn. Soc., Bot. 9: 154 (1865).—P. Taylor in Kew Bull. 18: 115 (1964); in Fl. Afr. Centr., Lentibulariaceae: 24 (1972); in F.T.E.A., Lentibulariaceae: 11 (1973). Type from S. Africa.
    Utricularia longecalcarata Benj. in Linnaea 20: 314 (1847).—Oliver in Journ. Linn. Soc., Bot. 9: 151 (1865).—Kam. in Engl., Bot. Jahrb. 33: 93 (1902). Type from S. Africa.
    Utricularia madagascariensis A. DC. in DC., Prodr. 8: 20 (1844). Type from Madagascar.
    Utricularia sanguinea Oliver in Journ. Linn. Soc., Bot. 9: 153 (1865).—Hiern, Cat. Afr. Pl. Welw. 1: 788 (1900).—Kam. in Engl., Bot. Jahrb. 33: 96 (1902).—S. Moore in Journ. Bot. 41: 405 (1903).—Stapf in Hook., Ic. Pl. 28, t. 2795A (1905).—Dyer, F.T.A. 4, 2: 475 (1906).—Burtt Davy & Pott-Leendertz, Check-list Fl. Pl. Transv.: 165 (1912).—Good in Journ. Bot. 68, Suppl. Gamopet.: 124 (1920). Type from Angola.
    Utricularia spartea Baker in Journ. Linn. Soc., Bot. 20: 216 (1883).—Vatke, Rel. Rutenb. 6: 130 (1885).—Kam. in Engl., Bot. Jahrb. 33: 94 (1902).—Perrier in Mém. Inst. Sci. Madag., Sér. B, 5: 195 (1955).—Humbert, Fl. Madag. Lentibulariaceae: 14 (1955). Type from Madagascar.
    Utricularia ibarensis Baker in Journ. Linn. Soc., Bot. 21: 427 (1885).—Kam. in Engl., Bot. Jahrb. 33: 100 (1902).—Perrier in Mém. Inst. Sci. Madag., Sér. B, 5: 196 (1955),.- Humbert, Fl. Madag. Lentibulariaceae: 15 (1955). Type from Madagascar.
    Utricularia dregei Kam. in Engl., Bot. Jahrb. 33: 94 (1902). Syntypes from S. Africa and Tanzania.
    Utricularia dregei var. *stricta* Kam., tom. cit.: 95 (1902). Syntypes from S. Africa.
    Utricularia engleri Kam., tom. cit.: 95 (1902). Syntypes from S. Africa.
    Utricularia livida var. *pauciflora* Kam., tom. cit.: 94 (1902). Type from S. Africa.
    Utricularia livida var. *micrantha* Kam., tom. cit.: 94 (1902). Type from S. Africa.
    Utricularia sanguinea var. *minor* Kam., tom. cit.: 96 (1902). Type from S. Africa.
    Utricularia elevata Kam., tom. cit.: 99 (1902). Syntypes from S. Africa.
    Utricularia elevata var. *macowanii* Kam., tom. cit.: 100 (1902). Type from S. Africa.
    Utricularia sprengelii var. *humilis* Kam., tom. cit.: 101 (1902). Type from Madagascar.

Utricularia prehensilis var. *huillensis* Kam., tom. cit.: 103 (1902), partim, quoad syn. *U. madagascariensis* A. DC.

Utricularia livida var. *engleri* (Kam.) Stapf in Dyer, F.C. 4, 2: 426 (1904).—Hook., Ic. Pl. 28, t. 2796/13—15 (1905).—Burtt Davy & Pott-Leendertz, Check-list Fl. Pl. Transv.: 164 (1912).

Utricularia tribracteata sensu Kam. in Engl., Bot. Jahrb. 33: 99 (1902), partim, excl. spec. Schimper 1943.—Stapf in Dyer, F.C. 4, 2: 427 (1904).—Hook., Ic. Pl. 28, quoad t. 2795B (1905), partim, excl. spec. Schimper 1943.—Dyer, F.T.A. 4, 2: 476 (1906), partim, excl. spec. Schimper 1943.—J.M. Wood in Trans. S. Afr. Phil. Soc. 18: 202 (1908).—Burtt Davy & Pott-Leendertz, Check-list Fl. Pl. Transv.: 165 (1912).— F.W. Andr., Fl. Pl. Sudan 3: 153 (1956), partim, quoad spec. "Imatong Mts.", non Hochst. ex A. Rich.

Utricularia transrugosa Stapf in Dyer, F.C. 4, 2 (1904); in Hook., Ic. Pl. 28, t. 2796B/16 & 17 (1905).—Dyer, F.T.A. 4, 2: 473 & 574 (1906).—Burtt Davy & Pott-Leendertz, Check-list Fl. Pl. Transv.: 165 (1912).—Suesseng. & Merxm. in Trans. Rhod. Sci. Ass. 43: 120 (1951).—Slinger in Bothalia 6: 355 (1954). Syntypes from S. Africa.

Utricularia kirkii sensu Stapf in Dyer, F.T.A. 4, 2: 476 (1906), partim, quoad spec. Kirk, "Batoka Country" & 574, quoad spec. Schimper 1943.—R.E. Fries in Wiss. Ergebn. Schwed. Rhod-Kongo-Exped. 1911—12, 1: 298 (1916).—Lloyd, Carniv. Pl. 232, 260, t. 33/1 & 2 (1942) (non Stapf in Dyer, F.C. 4, 2: 428 (1904)).

Utricularia odontosperma Stapf in Dyer, F.T.A. 4, 2: 474 (1906).—P. Taylor in Mem. N.Y. Bot. Gard. 9: 15 (1954). Syntypes from Malawi, Shire Highlands, *Buchanan* s.n. (K), near Blantyre, *Last* s.n. (K); Tanzania (or Zambia), between Lake Tanganyika and Lake Rukwa, *Nutt* s.n. (K).

Utricularia sematophora Stapf in Engl., Bot. Jahrb. 40: 60 (1907). Type from Tanzania.

Utricularia eburnea R.E. Fries, Wiss. Ergebn. Schwed. Rhod.-Kongo-Exped. 1911—12, 1: 297 (1916). Type from Zaire.

Utricularia humilis Phillips in Ann. S. Afr. Mus. 16: 230, t. 2/B (1917). Type from Lesotho.

Utricularia afromontana R.E. Fries in Notizbl. Bot. Gart. Berl. 8: 703 (1924). Type from Kenya.

Utricularia exilis sensu P. Taylor in Mem. N.Y. Bot. Gard. 9: 16 (1954) non Oliver.

Utricularia humbertiana Perrier in Mém. Inst. Sci. Madag., Sér. B, 5: 194 (1955); in Humbert, Fl. Madag. Lentibulariaceae: 11 (1955). Type from Madagascar.

Utricularia humbertiana var. *andringitrensis* Perrier, tom. cit.: 195 (1955); in Humbert, loc. cit.: 14 (1955), pro syn.

Utricularia spartea var. *marojejensis* Perrier, tom. cit.: 196 (1955); in Humbert, tom. cit.: 15 (1955). Type from Madagascar.

Utricularia spartea var. *subspicata* Perrier, tom. cit.: 196 (1955); in Humbert, tom. cit.: 15 (1955). Syntypes from Madagascar.

Utricularia parkeri sensu Perrier in Humbert, tom. cit.: 16 (1955), partim, quoad syn. *U. sprengelii* var. *humilis* Kam. non Baker.

Utricularia mauroyae Perrier, tom. cit.: 200 (1955); in Humbert, tom. cit.: 12 (1955). Type from Madagascar.

Terrestrial herb. Rhizoids and stolons capillary, usually numerous from the base of the peduncle. Leaves not always present nor conspicuous at anthesis, subrosulate at the peduncle base and scattered on the stolons, linear to obovate-spathulate or rarely reniform, 1—7 cm. long, 1—6 mm. wide; nerves of the lamina usually dichotomously branched. Traps numerous, ovoid, 1—2 mm. long, stalked; mouth terminal, provided with dorsal and ventral radiating rows of glandular hairs. Inflorescence erect, straight or flexuous, simple or rarely branched above, 2—80 cm. high; smooth and glabrous, relatively robust; flowers (1)2—8(50), distant or more rarely congested; scales few, similar to the bracts; bracts basifixed, ovate, acute or acuminate,more or less 1 mm. long; bracteoles linear-lanceolate, more or less as long as the bract; pedicels 0.5—1(3) mm. long, erect at anthesis, spreading or deflexed in fruit. Calyx lobes subequal, ovate, 2—3 mm. long at anthesis, accrescent, always plicate along the nerves, with apex of the upper lobe acute or obtuse, of lower lobe rounded, truncate or more or less bidentate. Corolla violet, mauve or white with a yellow spot on the palate or more rarely wholly yellow or cream, 5—15 mm. long; superior lip 1.5—2 times as long as the upper calyx lobe, narrowly oblong-ovate, with apex rounded or truncate; inferior lip circular; palate raised and double crested, the crests usually transversely tuberculate; spur slightly shorter than to about 1.5 times as long as the inferior lip, conical-subulate, straight or curved. Filaments linear; anther-

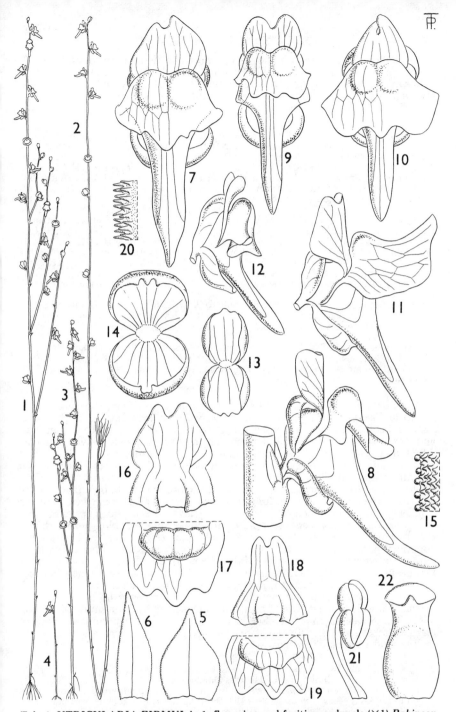

Tab. 1. UTRICULARIA FIRMULA. 1, flowering and fruiting peduncle (×1) *Robinson* 760; 2, flowering and fruiting peduncle (×1) *Milne-Redhead & Taylor* 10581; 3, flowering and fruiting peduncle (×1), *Milne-Redhead & Taylor* 9882; 4, flowering peduncle (×1), *Fanshawe* 85; 5, bract (×30), 6, bracteole (×30), 5—6 from *Hope-Simpson* 102; 7, flower, abaxial view (×12), 8, flower, lateral view (×12), 7—8 from *Milne-Redhead & Taylor* 8059a; 9, flower, abaxial view (×12), *Jackson* 1729; 10, flower, abaxial view (×12), *Bogdan* 4662; 11, flower, lateral view (×12), *Richards* 9933e; 12, flower, lateral view (×12); 13, calyx, abaxial view (×12), 12—13 from *Harley* 1853; 14, calyx, abaxial view (×12), *Richards* 9933e; 15, calyx margin (×150), *Milne-Redhead & Taylor* 8059a; 16, superior corolla lip (×12), *Milne-Redhead & Taylor* 8059; 17, superior corolla lip (×12), *Harley* 1853; 18, inferior corolla lip (×12), *Milne-Redhead & Taylor* 8059; 19, inferior corolla lip (×12), *Harley* 1853; 20, papillae on spur (×150), *Bogdan* 4662; 21, stamen (×30); 22, pistil (×3), 21—22 from *Jackson* 1729.

thecae subdistinct. Ovary globose; style short but distinct; stigma inferior lip semi-circular, superior much smaller, deltoid. Capsule globose, up to 2 mm. long, dehiscing by dorsal and ventral longitudinal slits with thickened margins. Seed few to many, 0.3−0.5 mm. long, ovoid, slightly angular, smooth or obscurely to distinctly papillose; testa-cells indistinct, elongate.

Zambia. N: Chinsali Distr., along chansh R., W. of Shiwa Ngandu Estates on Rd. to Kasama, fl. 28.viii.1979, *Chisumpa* 612 (K; NDO). W: Mwinilunga Distr., 100 m. NE. of Dobeka Br., fl. 11.xii.1937, *Milne-Redhead* 3611 (K). C: Serenje, fl. 27.ix.1961, *Fanshawe* 6719 (K; NDO). E: Chipata Distr., Lunkwakwa, fl. 12.x.1967, *Mutimushi* 2138 (K; NDO). **Zimbabwe.** N: Mazoe, Umvukwes, Ruorka Ranche, 1666 m., fl. 17.xii.1952, *Wild* 3933 (K; MO; SRGH). W: Matobo, Farm Besna Kobila, 1660 m., fl. viii.1957, *Miller* 4514 (K; SRGH). C: Harare, Mansala R., Shamva Rd., fl. 24.x.1960, *Rutherford-Smith* 340 (K; SRGH). E: Inyanga, Mt. Dombo, 636 m., fl. 17.vii.1974, *Müller* 2176 (K; SRGH). S: Livingstone Isl., Victoria Falls, 1000 m., fl. ix.1905, *Gibbs* 176 (BM). **Malawi.** N: Mzimba Distr., 4.8 km. SW. of Mzuzu, 1500 m., fl. 11.i.1976, *Pawek* 10700 (K; MAL; MO; SRGH; UC). S: Zomba Distr., Zomba Mt., 1490 m., fl. 31.iii.1978, *Pawek* 14164 (K; MAL; MO). **Mozambique.** N: Niassa, 36 km. E. of Ribáuè, 666 m., fl. 17.v.1961, *Leach & Rutherford-Smith* 10911 (K; SRGH). Z: Gúruè, Haute Vallee du Rio Malema base de la Serra Namuli, 1330 m., fl. 2.viii.1979, *Schafer & Koning* 6931 (K; LIJSC). MS: Beira, Cheringoma Coast, Nyamaruzu Dambo, halfway between Camp and Rd. junction to Chinziua Lighthouse, fl. v.1973, *Tinley* 2908 (K; SRGH).

Widespread in eastern Africa from Ethiopia to S. Africa (Cape Prov.), Madagascar and also in Mexico. In permanently or seasonally wet boggy grassland and in shallow wet soil over rocks; sea level to 2600 m.

4. **Utricularia pentadactyla** P. Taylor in Mem. N.Y. Bot. Gard. 9: 16 (1954); in Kew Bull. 18: 129 (1964); in Fl. Afr. Centr. Lentibulariaceae: 14 (1972); in F.T.E.A., Lentibulariaceae: 9 (1973). Type: Zimbabwe, Harare, Dombashawa, *Wild* 3240 (K, holotype; SRGH, isotype).

Terrestrial herb. Rhizoids and stolons capillary, few from the base of the peduncle. Leaves often decayed at anthesis, or 2−5 subrosulate at the peduncle-base and scattered on the stolons, oblanceolate-spathulate, up to 6 mm. long and 1 mm. wide, 1-nerved. Traps numerous on the rhizoids and leaves, globose, 0.6−0.8 mm. long, shortly stalked; mouth terminal , provided with dorsal and ventral radiating comb-like rows of glandular hairs. Inflorescence erect, 2−30 cm. high; peduncle filiform, glabrous above, papillose or setulose at the base; flowers 1−4, distant; scales few, similar to the bracts; bracts basifixed, ovate-deltoid, more or less 0.6 mm. long; bracteoles similar but narrower; pedicels capillary, erect, shorter than the calyx-lobes. Calyx lobes unequal; upper broadly ovate, 1−1.6 mm. long, obtuse to subacute; lower smaller, oblong, emarginate. Corolla pale mauve to white with a yellow spot on the palate, 3−15 mm. long; superior lip narrowly oblong, 1.5−2.5 times as long as the upper calyx lobe, with apex emarginate to deeply bifid; inferior lip circular in outline, more or less deeply 5-lobed palate raised, often crested; spur subulate, slightly curved, up to 10 mm. long. Filaments filiform; anther-thecae subdistinct. Ovary ovoid; style very short; stigma inferior lip semi-circular, superior lip much smaller, deltoid. Capsule globose, 1.5−2 mm. long, dehisceing by a longitudinal ventral slit with thickened margins. Seeds numerous, somewhat conical, slightly angular; testa smooth, cells indistinct, elongated on the lateral surfaces, more or less isodiametric elsewhere.

Zambia. N: Mbala Distr., Kali Dambo near Kawimbi Mission, 1700 m., fl. 6.v.1952, *Richards* 1635A (K). **Zimbabwe.** C: Harare, Dombashawa, 1666 m., fl. 8.iii.1950, *Wild* 3240 (K, type; SRGH, isotype). **Malawi.** N: Rumphi Distr., Nyika Plateau, Chowo Rocks, 2210 m., fl. 17.v.1970, *Brummitt* 10867 (K). S: Mulanje Distr., Mulanje Mt., path from Thuchila Hut to head of Ruo basin, 1980 m., fl. 8.iv.1970, *Brummitt* 9770 (K).

Also in Zaire, Ethiopia, Uganda, Kenya, Tanzania and Angola. Damp sandy or peaty grassland and in shallow wet soil over rocks; 1500−2100 m.

5. **Utricularia welwitschii** Oliver in Journ. Linn. Soc., Bot. 9: 152 (1865).−Stapf in F.T.A. 4, 2: 478 (1906).−P. Taylor in Kew Bull. 18: 144 (1964), pro parte, excl. vars.; in Fl. Afr. Centr., Lentibulariaceae: 18, t. 4/1−18 (1972); in F.T.E.A., Lentibulariaceae: 10 (1973). Type from Angola.

Utricularia welwitschii var. welwitschii P. Taylor in Kew Bull. 18: 146 (1964). Type from Angola?

Terrestrial herb. Rhizoids and stolons capillary, numerous from the base of the peduncle. Leaves scattered on the stolons, often decayed at anthesis, obovate-spathulate to linear-oblanceolate, up to 30 × 0.5−2 mm., 1-nerved. Traps numerous, ovoid, 0.4−0.8 mm. long, stalked; mouth terminal with a dorsal fringe of more or less 5 glandular hairs and ventral radiating rows of more or less sessile glands. Inflorescence erect, simple or branched above, 4−50 cm. high; peduncle usually robust, straight or twining, smooth; inflorescence-axis usually flexuous; flowers 1−25 or more, usually distant; scales numerous, similar to the bracts; bracts basifixed, lanceolate to ovate, 0.6−1.6 mm. long; bracteoles lanceolate, more or less as long as the bracts; pedicels up to 1 mm. long. Calyx lobes subequal, circular, 1.5−2.5 mm. long, densely papillose, margin strongly incurved, with apex tridenticulate. Corolla usually violet with a yellow blotch on the palate, sometimes wholly yellow, 5−15 mm. long; superior lip broadly ovate to circular, 2−4 times as long as the upper calyx lobe, with apex rounded, truncate or emarginate; inferior lip circular, with apex rounded, entire or obscurely 3-crenate; palate raised, usually distinctly 2-gibbous; spur subulate, acute, 1.5−2 times as long as the inferior lip. Filaments filiform; anther-thecae subdistinct. Ovary globose; style obsolete; stigma inferior lip semi-circular, superior much smaller, deltoid. Capsule globose, more or less 1.5 mm. long, dehiscing by dorsal and ventral ovate-lanceolate pores. Seeds numerous, ovoid, angular, 0.25 mm. long; testa smooth, cells indistinct, more or less isodiametric.

Zambia. B: 32 km. NE. of Mongu Distr., fl. 10.xi.1959, *Drummond & Cookson 6302* (K; SRGH). N: Mansa, Chimana Dambo, near Samfya Mission on Lake Bangweulu, fl. 8.x.1947, *Brenan & Greenway 8076* (FHO; K). W: Mwinilunga Distr., Zambesi R., 6.5 km. N. of Kalene Hill Mission, fl. 16.vi.1963, *Edwards 803* (K; SRGH). C: Serenje, fl. 27.ix.1961, *Fanshawe 6718* (EA; K). Zimbabwe. N: Umvukwe Hills, fl. 5.vii.1934, *Gilliland Q622* (BM; K). W: Matobo, Farm Besna Kobila, 1600 m., fl. ix.1957, *Miller 4552* (BR; K; SRGH). C: Harare, Ruwa R., 1500 m., fl. 23.xi.1947, *Wild 2254* (BR; K; SRGH). E: Inyanga, *Gilliland K1079* (BM; K; SRGH). Malawi. N: Nyika Plaeau, Lake Kaulime, 2200 m., fl. 23.x.1958, *Robson 285* (K; LISC; SRGH).
Also known from Rwanda, Burundi, Central African Republic, Tanzania, Angola, S. Africa (Transvaal) and Madagascar. Damp sandy or peaty grassland; 900−2000 m.

6. **Utricularia odontosepala** Stapf in Kew Bull. **1912**: 331 (1912).−P. Taylor in Fl. Afr. Centr., Lentibulariaceae: 20 (1972). Type: Zambia, Kashitu R., *Rogers 8632* (K, holotype; GRA, isotype).
Utricularia welwitschii var. *odontosepala* (Stapf) P. Taylor in Kew Bull. 18: 148 (1964). Type as above.

Terrestrial herb. Rhizoids and stolons capillary, numerous at the base of the peduncle. Leaves scattered on the stolons, narrowly obovate-spathulate, 10−30 × c. 1 mm., 1-nerved. Traps numerous on the stolons and leaves, ovoid, 0.3-0.8 mm. long, stalked; mouth terminal, with a dorsal fringe of c. 5 glandular hairs and ventral radiating rows of more or less sessile glands. Inflorescence erect, simple, 5−25 cm. high; peduncle filiform, straight, glabrous above, setulose at the base; inflorescence axis straight; flowers 1−5, usually distant; scales numerous, similar to the bracts; bracts basifixed, lanceolate with apex acute, 1.2−1.5 mm. long; bracteoles similar to the bracts but slightly longer; pedicels up to 1 mm. long. Calyx lobes subequal, broadly ovate, 2.5−3.5 mm. long, minutely papillose or smooth, flat, the margin deeply divided into 7−11 subulate teeth. Corolla 7−15 mm. long, mauve or violet with a yellow spot on the palate; superior lip broadly ovate, with apex retuse, 2−25 times as long as the upper calyx lobe; inferior lip reniform with apex distinctly 3-crenate; palate raised, 2−4 gibbous; spur subulate, acute, about 1.5 times as long as the inferior lip. Filaments filiform; anther thecae sub-distinct. Ovary globose; style very short; stigma inferior lip semi-circular, superior much smaller, deltoid. Capsule globose, 1.5−2 mm. long, dehiscing by dorsal and ventral lanceolate pores. Seeds numerous, ovoid, somewhat angular, c. 0.2 mm. long; testa smooth, the cells indistinct, more or less isodiametric.

Zambia. W: Chizela, fl. 11.vi.1953, *Fanshawe* 80 (BR; EA; K; SRGH). C: Kashitu R.,
fl. vii.1905, 1400 m., *Rogers* 8632 (GRA; K, type). **Malawi**. N: Rumphi Distr., Nyika Nat.
Park, Chowo Rock, 2366 m., fl. 28.iv.1973, *Pawek* 6670 (CAH; K; MO; UC; UM?).
Also in Katanga. Damp grassland and shallow wet soil over rocks; 1300−2200 m.

7. **Utricularia microcalyx** P. Taylor in Bull. Jard. Bot. Nat. Belg. **41**: 270 (1971); in Fl.
Afr. Centr., Lentibulariaceae: 19 (1972). Type: Zambia, Kambole Escarpment,
Richards 10051 (BR; K, holotype; EA; LISC; PRE; S; SRGH, isotypes).
  Utricularia welwitschii var. *microcalyx* P. Taylor in Kew Bull. **18**: 148 (1964). Type
as above.

Terrestrial herb. Rhizoids and stolons capillary, numerous at the base of the
peduncle. Leaves scattered on the stolons, obovate-spathulate, c. 10 × 1 mm.,
1-nerved. Traps numerous on the stolons and leaves, ovoid, 0.4−0−8 mm. long,
stalked; mouth terminal with a dorsal fringe of c. 5 glandular hairs. Inflorescence
erect, simple, 5−30 cm. high; peduncle filiform, straight, glabrous above,
minutely papillose at the base; inflorescence axis straight; flowers 1−7, distant;
scales numerous, similar to the bracts; bracts basifixed, ovate with apex acute,
up to 0.8 mm. long; bracteoles narrowly oblong with apex obtuse, about as long
as the bracts; pedicels up to 1 mm. long. Calyx lobes unequal, flat, minutely
papillose, the upper lobe circular with apex emarginate, c. 1.5 mm. long, the
lower lobe much smaller, ovate or oblong with apex emarginate. Corolla mauve
or pale blue with a yellow spot on the palate, 7−14 mm. long; superior lip broadly
ovate with apex truncate, 3−5 times as long as the upper calyx lobe; inferior
lip reniform with apex distinctly 3-crenate; palate raised, distinctly 4-gibbous;
spur subulate, acute, about 1.5 times as long as the inferior lip. Filaments
filiform; anther thecae sub-distinct. Ovary ovoid; style more or less obsolete;
stigma inferior lip semi-circular, superior lip much smaller, oblong. Capsule
globose, 1.5−2 mm. long, dehiscing by dorsal and ventral lanceolate pores. Seeds
numerous, ovoid, somewhat angular, c. 0.2 mm. long; testa smooth, the cells
indistinct, more or less isodiametric.

Zambia. N: Mbala Distr., Kambole Escarpment near Ngozi R., 1500 m., fl. 6.vi.1957,
Richards 10051 (BR; EA; K, type; LISC; PRE; S; SRGH).
Also known from Katanga.

8. **Utricularia firmula** Oliver in Journ. Linn. Soc., Bot. **9**: 152 (1865).—Stapf in Dyer,
F.T.E.A. **4**: 479 (1906).—P. Taylor in F.W.T.A., ed. 2, **2**: 378 (1963); in Kew Bull.
18: 151 (1964); in Fl. Afr. Centr., Lentibulariaceae: 16 (1972); in F.T.E.A.,
Lentibulariaceae: 9 (1973). TAB. 1. Type from Angola.
  Utricularia ecklonii var. *lutea* Perrier in Mém. Inst. Sci. Madag., Sér. B. **5**: 198
(1955). Type from Madagascar.

Terrestrial herb. Rhizoids and stolons capillary, numerous from the base of
the peduncle. Leaves scattered on the stolons, narrowly obovate-spathulate,
2−20 × more or less 1 mm. wide, 1-nerved. Traps numerous, ovoid, 0.15−0.3
mm. long, stalked, the stalk more or less as long as the trap; mouth terminal,
dorsally fringed with 2−10 glandular hairs. Inflorescence stiffly erect, simple
or sometimes branched above, 2, 2−36 cm. high; peduncle relatively thick and
rigid, glabrous; flowers (1)10−20(50); scales numerous, especially below, similar
to the bracts; bracts basifixed, ovate-deltoid, reflexed; bracteoles lanceolate,
erect; peduncles erect, 0.2−0.5 mm. long. Calyx lobes subequal, circular, very
concave with apex emarginate or shortly 2−3-denticulate. Corolla pale yellow,
3.5−6 mm. long, persistent; superior lip broadly ovate with apex emarginate,
more or less 1.5 times as long as the upper calyx lobe; inferior lip longer, circular
in outline, distinctly 3-lobed; palate slightly raised; spur subulate, 2.5−3.5 times
as long as the inferior lip. Filaments filiform, curved; anther-thecae more or
less confluent. Ovary ovoid; style very short; stigma inferior lip rounded,
superior lip minute, deltoid. Capsule globose, more or less 1.2 mm. long,
dehiscing by dorsal and ventral ovate-lanceolate pores. Seeds numerous,
narrowly truncate-conical; testa smooth, cells indistinct, elongate.

Zambia. B: Kaoma, near Resthouse, fl. 20.xi.1959, *Drummond & Cookson* 6674 (K;
SRGH). N: Mbala Distr., Kambole Escarpment near Ngozi R., 1500 m., fl. 6.vi.1957,
Richards 10043 (BR; EA; K; M; PRE; S; SRGH; W; Z). W: Chizela, fl. 11.vi.1953, *Fanshawe*

76 (K). C: Mkushi Distr., near Kantwite R., on old Mkushi Rd., 43 km. E. of Kabwe, 1160 m., fl. 20.iv.1972, *Kornas* 1629 (K). S: Siamambo Forest Reserve near Choma, fl. 23.vii.1952, *Angus* 18 (FHO; K). **Zimbabwe**. W: Matobo, Farm Besna Kobila, 1600 m., fl. viii.1957, Miller 4513 (K; SRGH). C: Harare, Mutare Rd., fl. 18.x.1969, Rutherford-Smith (K; SRGH). S: Bikita, Turgwe-Dafana confluence, 1050 m., fl. 5.v.1969, *Biegel* 3018 (K; SRGH). **Malawi**. N: Mzimba Distr., 4.8 km. W. of Mzuzu, Katoto, 1500 m., 23.vii.1973, *Pawek* 7224 (K; MAL; MO; SRGH; UC). C: Lilongwe, Dzalanyama Forest Reserve, above Chaulongwe Falls, fl. 26.iv.1970, *Brummitt* 10174 (K). S: Zomba Distr., Chindola, fl. 15.vii.1955, *Jackson* 1729 (K). **Mozambique**. Z: entre Maganja da Costa e o Regulo Ingwie a 19 km. de Maganja fl. 26.ix.1949, *Barbosa & Carvalho* 4193 (K; LMJ). MS: 36.8 km. from Beira, fl. 17.viii.1908, *Johnson* 295 (K). GI: Bilene, 1.6 km., fl. 25.ix.1978, *Schäfer* 6537 (K; LMU). M: Inyanasan, fl. 22.i.1898, *Schlechter* 12070 (K).

Widespread in Africa from Senegal to Sudan, Angola, S. Africa (Natal), and Madagascar. Boggy grassland and wet rocks; sea level to 2100 m.

9. **Utricularia pubescens** Sm. in Rees., Cyclop. 37, no.53 (1819).—P. Taylor in F.W.T.A., ed. 2, 2: 378 (1963); in Kew Bull. **18**: 101 (1964); in Fl. Afr. Centr., Lentibulariaceae: 29 (1972); in F.T.E.A., Lentibulariaceae: 12 (1973). Type from Sierra Leone.
   Utricularia papillosa Stapf in Bull. Misc. Inf. Kew 1916: 41 (1916).—Hutch. & Dalz., F.W.T.A. 2: 234 (1931). Syntypes from Nigeria.
   Utricularia graniticola A. Chev. & Pellegrin in Mém. Soc. Bot. Fr. 8: 276 (1917).—Hutch. & Dalz., F.W.T.A. 2: 234 (1931). Syntypes from Ivory coast.
   Utricularia peltatifolia A. Chev. & Pellegrin in loc. cit. Syntypes from Guinee, Ivory Coast & Katanga.
   Utricularia hydrocotyloides Lloyd & G. Taylor in Contr. Gray Herb. **165**: 85 (1947). Type from Angola.
   Utricularia fernaldiana Lloyd & G. Taylor in tom. cit. (1947). Type from Uganda.
   Utricularia thomasii Lloyd & G. Taylor in tom. cit. (1947). Type from Uganda.
   Utricularia deightonii Lloyd & G. Taylor in tom. cit. (1947). Type from Sierra Leone.

Terrestrial herb. Rhizoids and stolons capillary, numerous from the base of the peduncle. Leaves usually present at anthesis, scattered on the stolons, circular, peltate, long-petiolate; lamina fleshy, mucilaginous, 1—5 mm. in diam.; petiole 2—10 mm. long. Traps numerous, globose, stalked, 0.5—0.8 mm. long, mouth terminal, with dorsal and ventral radiating rows of glandular hairs. Inflorescence erect, 2—35 cm. high; peduncle filiform, straight, more or less setulose, papillose or rarely entirely glabrous; flowers 1—6(10), distant; scales few, similar to the bracts; bracts basisolute, shortly produced below the point of insertion, ovate, with apex acute, base truncate, more or less 1 mm. long, setulose; bracteoles similar to the bracts but narrower; pedicels erect, 0.5—2 mm. long. Calyx lobes subequal, usually setulose; upper lobe ovate, 1.5—3 mm. long, with apex acute; lower lobe ovate or circular, with apex bidentate. Corolla white or pale lilac, 2—16 mm. long; superior lip 1.5—2 times as long as the upper calyx lobe, narrowly oblong with apex rounded, truncate or more or less emarginate; inferior lip circular, entire; palate much raised, double crested, the crests usually transversely tuberculate; spur usually conical-subulate, up to 4 times as long as the inferior lip but very variable in length and sometimes much reduced. Filaments filiform; anther-thecae subdistinct. Ovary ovoid; style short but usually distinct; inferior lip of stigma semi-circular, superior much smaller, deltoid. Capsule globose dehiscing by a longitudinal ventral slit with thickened margins. Seeds numerous, ovoid; testa-cells distinct, more or less isodiametric.

Zambia. N: Mansa, fl. 8.x.1947, *Brenan & Greenway* 8077 (EAH; FHO; K). W: Mwinilunga Distr., Sinkabolo Swamp, 1200 m., fl. 20.xi.1962, *Richards* 7450 (K). **Malawi**. N: Rumphi Distr., Nyika Plateau, 12.8 km. N. of M1, fl. 11.iii.1978, *Pawek* 14074 (K; MAL; MO).

Widespread in Africa from Guineé to Ethiopia and Angola, in S. America and in India. Damp peaty soil in wet grassland and on wet rocks; sea level to 1900 m.

10. **Utricularia podadena** P. Taylor in Kew Bull. **18**: 78 (1964). Type: Malawi, Zomba Distr., Chindola Dambo, *Jackson* 1728 (K, holotype).

Terrestrial herb. Rhizoids and stolons numerous from the base of the peduncle. Leaves few, scattered on the stolons, obovate-spathulate, up to 6 × 1 mm., 1-nerved. Traps few, stalked, ovoid, c. 0.3 mm. long; mouth lateral with a single reflexed subulate, apically shortly bifid dorsal appendage and 2 much shorter subulate ventral appendages. Inflorescence erect, up to 24 cm. tall; peduncle filiform, sparsely covered (as are the bracts, bracteoles, pedicels, calyx and spur of corolla) with multicellular gland-tipped hairs; flowers 3−9, distant; scales few, similar to the bracts; bracts basifixed, ovate with apex subacute, c. 0.7 mm. long; bracteoles basifixed, narrowly oblong, about as long as the bracts; pedicels filiform, spreading at anthesis, deflexed in fruit, up to 7 mm. long. Calyx lobes unequal; upper lobe quadrate with apex rounded, c. 2 mm. long; lower lobe about as long, narrowly oblong with apex emarginate. Corolla yellow, c. 8 mm. long; superior lip transversely oblong, 2−3 times as long as the upper calyx lobe; inferior lip transversely elliptic with apex obscurely 3-crenate; palate slightly gibbous, obscurely longitudinally 4−6 ridged; spur straight or slightly curved, subulate, about as long as and widely diverging from the inferior lip. Filaments linear; anther-thecae subdistinct. Ovary ovoid, glandular above, style very short, narrowly winged above; stigma inferior lip circular; superior lip much smaller, deltoid. Capsule globose, 1.5−2 mm. long, dehiscing by a ventral, longitudinal, marginally thickened slit. Seeds numerous, ovoid, obscurely reticulate; testa cells elongate, irregular.

Malawi. S: Zomba Distr., Chindola Dambo, fl. 15.vii.1955, *Jackson* 1728 (K, type).
Mozambique. N: Lichinga, 300 m., fl. 1.vii.1934, *Torre* 144 (COI; LISC).
Known only from Malawi and Mozambique. In marshy grassland; c. 1000 m.

11. Utricularia simulans Pilger in Not. Bot. Gart. Berl. 6: 194 (1914). Type from Brazil.
  Utricularia fimbriata sensu Taylor in Kew Bull. 18: 72 (1964); in Fl. Afr. Centr., Lentibulariaceae: 26 (1972) non Kunth.

Terrestrial herb. Rhizoids and stolons capillary, numerous at the base of the peduncle. Leaves numerous, rosulate at the base of the peduncle and scattered on the stolons, narrowly linear, 5−15 × 0.3−0.5 mm., 1-nerved. Traps numerous, stalked, or 0.2−0.3 mm. long; mouth lateral with a shortly conical dorsal appendage and a larger, deeply bifid ventral appendage. Inflorescence erect, 1.5−11 cm. tall; peduncle filiform, glabrous; flowers 1−5, more or less congested; scales numerous, similar to the bracts; bracts basifixed, broadly ovate in outline, 1−1.5 mm. long, the base auriculate, the margin deeply divided into up to 15 subulate acute teeth; bracteoles inserted on the first half of the pedicel, much longer than the bracts, circular in outline, narrowed at the base, c. 3 mm. long, the margin deeply divided into c. 11 subulate acute teeth; pedicels filiform, erect 0.5−1 mm. long. Calyx lobes subequal, 4−5 mm. long, circular in outline, the upper lobe with margin deeply divided into c.15 subulate acute teeth, the lower lobe slightly larger with apex bifid and margin divided into c. 18 subulate acute teeth. Corolla yellow, 5−10 mm. long; superior lip oblong with apex truncate or emarginate; inferior lip circular or broadly ovate; palate slightly raised; spur narrowly conical with apex obtuse or subacute, about as long as the inferior lip. Filaments linear, anther thecae more or less confluent. Ovary ovoid; style distinct; stigma inferior lip semi-circular; superior lip much smaller, deltoid. Capsule (not seen in African specimens) globose, c. 2 mm. long, dehiscing by a longitudinal ventral, marginally thickened slit. Seeds numerous, ovoid, c. 1.5 mm. long; testa thin, reticulate, the cells elongate.

Zambia. N: Mpika Distr., Serenje-Mpika Rd., fl. 5.iv.1961 *Richards* 14990 (K).
Also in Senegal, Guinea Bissau, Mali, Liberia, Cameroun, Chad, Gabon, Zaire, Angola and America from Florida to southern Brazil, Bolivia and Paraguay. Wet sandy savanna; from sea level to 1575 m.

12. Utricularia andongensis Hiern, Cat. Afr. Pl. Welw. 1: 787 (1900).—Stapf in Dyer, F.T.A., 4: 481 (1906).—P. Taylor in F.W.T.A., ed. 2: 377 (1963); in Kew Bull. 18: 38 (1964); in Fl. Afr. Centr., Lentibulariaceae: 8 (1972); in F.T.E.A., Lentibulariaceae: 6 (1973). Type from Angola.
  Utricularia prehensilis var. *parviflora* Oliver in Journ. Linn. Soc., Bot. 9: 150 (1865). Syntypes from Gabon and Angola.
  Utricularia tortilis var. *andongensis* (Hiern) Kam. in Engl., Bot. Jahrb. 33: 104 (1902) partly excl. spec. Gillet 2. Syntypes as above.
  Utricularia conferta sensu Kam. in Engl., Bot. Jahrb. 33: 102 (1902) non Wight.

Terrestrial herb. Rhizoids and stolons capillary, few from the base of the peduncle. Leaves 3 — 6, rosulate at the base of the peduncle and usually present at anthesis, linear, 3-nerved, up to 6 cm. long, 1.5 — 5 mm. wide. Traps few, globose, shortly stalked, glandular, 0.8 — 1.1 mm. long; mouth basal, with 2 dorsal subulate appendages. Inflorescence erect or rarely twining, 2 — 15(20) cm. high; filiform, glabrous; flowers 1 — 8; scales few, similar to the bracts; bracts basifixed, ovate, acuminate, 1.5 — 2 mm. long 1 — 3-nerved; bracteoles linear-subulate, shorter than the bracts; pedicels erect or ascending, (1)1.5 — 13 times as long as the calyx, narrowly winged. Calyx lobes broadly ovate, 1.5 — 2.5 mm. long at anthesis, becoming longer and relatively broader in friut, slightly unequal; upper acute; lower shorter, with apex bidentate. Corolla yellow, 4 — 10 mm. long; superior lip more or less as long as the upper calyx-lobe, narrowly oblong with apex rounded, truncate or emarginate; inferior lip circular, with apex entire, emarginate or obscurely 3-crenate; palate scarcely raised, obscurely 2-gibbous; spur conical-subulate, acute, curved. Filaments linear, slightly curved; anther thecae subdistinct. Ovary narrowly ovoid; stigma subsessile, inferior lip truncate or rounded, superior lip smaller, truncate. Capsule broadly oblong, dorsiventrally compressed, smaller than the calyx, dehiscing by a longitudinal ventral slit with thickened margins in the upper half. Seeds few, ovoid, verrucose, 0.4 — 0.6 mm. long; testa rather loose and corky, cells irregular, much longer than broad.

Zambia. W: Mwinilunga Distr., Slope of Matonchi Farm fl., 13.ii.1938, *Milne-Redhead* 4555 (K).
Widespread in Africa from Guineé to Angola. Wet moss-covered rocks, *Sphagnum* bogs and marshes; 240 — 1800 m.

13. **Utricularia scandens** Benj. in Linnaea 20: 309. Type from India.
 Utricularia gibbsiae Stapf in Dyer, F.T.A., 4: 574 (1906). Type: Zimbabwe, Victoria Falls, *Gibbs* 177 (K, holotype, BM, isotype).
 Utricularia schweinfurthii Stapf in Dyer, F.T.A., 4: 482 (1906). Type from Sudan.
 Utricularia scandens subsp. *scandens* P. Taylor in Kew Bull. 18: (1964); in Fl. Afr. Centr., Lentibulariaceae: 7 (1972); in F.T.E.A., Lentibulariaceae: 6 (1973). Type as for *Utricularia scandens*.
 Utricularia scandens subsp. *schweinfurthii* (Stapf) P. Taylor in F.W.T.A., ed. 2, 2: 378 (Oct 1963); in Kew Bull. 18: 48 (1964); in Fl. Afr. Centr., Lentibulariaceae: 8 (1972). Type as for *U. schweinfurthii*.
 Polypomphlyx madecassa Perrier in Mém. Inst. Sci. Madag., Sér. B, 5: 199 (1955). Type from Madagascar.

Terrestrial herb. Rhizoids and stolons capillary, few from the base of the peduncle. Leaves few, branch-opposed on the stolons, linear, 1-nerved, up to 10 × 1 mm. Traps scattered on the stolons and leaves, globose, shortly stalked, glandular, 0.6 — 1.0 mm. long; mouth basal, with 2 dorsal simple subulate appendages, a single shorter truncate or shortly bifid ventral appendage. Inflorescence erect or twining, 3 — 35 cm. high; peduncle filiform, glabrous; flowers 1 — 8, distant, usually alternating with sterile bracts; scales few, similar to the bracts; bracts basifixed, broadly ovate-deltoid, acute or acuminate, nerveless, 1 — 1.5 mm. long; bracteoles linear-lanceolate, more or less 1 mm. long; pedicels spreading or erect, 1 — 3 times as long as the flowering calyx, narrowly winged. Calyx lobes broadly ovate. 2.5 — 3 mm. long at anthesis, up to 5 mm. long in fruit, deccurent, slightly unequal; upper longer, with apex acute or acuminate; lower with apex shortly 2 — 3-dentate. Corolla yellow, 5 — 15 mm. long; superior lip shorter than to about twice as long as the upper calyx lobe, oblong to circular, with apex more or less emarginate; inferior lip circular, with apex rounded, entire or 2 — 3-crenate; palate raised, usually 2 — 4 gibbous; spur subulate, acute, more or less curved. Filaments linear, more or less straight; anther-theacae confluent. Ovary ovoid; style short; stigma-lips semi-circular, the superior smaller. Capsule oblong-ovoid, dorsiventrally compressed, 2 — 2.5 mm. long, dehiscing by a longitudinal ventral slit with thickened margins. Seeds ovoid or elliptic, smooth, more or less 0.2 mm. long; hilum prominent; testa-cells distinct, considerably longer than broad.

Botswana. N: E. end of Xanenga Isl., fl. 2.x.1975, *Smith* 1461 (K). **Zambia.** B: Kaoma,

near Resthouse, fl. 20.xi.1959, *Drummond & Cookson* 6675 (K; SRGH). N: Mbala, Kalambo Farm, fl. 21.v.1952, *Richards* s.n. (K). W: Mwinilunga Distr., fl. 8.x.1937, *Milne-Redhead* 2659 (K). **Zimbabwe**. W: Victoria Falls, fl. 27.iii.1906, *Gibbs* 177 (BM; K, type). C: Harare, Rumani 1500 m., fl. 10.ix.1952, *Wild* 3866 (K; SRGH). E: Chimanimani, Tarka Forest Reserve, N. bank of Chisengu R., fl. x.1968, *Goldsmith* 141/68 (K; SRGH). **Malawi**. N: Mzimba, 3 km. SW. of Mzuzu at Kototo (Bishop's House), fl. 8.iii.1977, *Grosvenor & Renz* 1077 (K; SRGH). **Mozambique**. MS: Base of Bandula Peak, fl., 4.iv.1952, c. 720 m., *Chase* 4453 (BM; SRGH). M: Namaacha, fl. 30.vi.1961, *Balsinhas* 506 (K; Centr. Invest. Agric.).
    Widespread in Africa from Guineé to S. Africa (Transvaal) and in Madagascar. Also in tropical Asia from India to New Guinea. Boggy grassland or among mosses on wet rocks; 1000–2250 m.

14. **Utricularia prehensilis** E. Mey., Comm. Pl. Afr. Austr.: 282 (1837).—Stapf in Dyer, F.T.A. 4: 480 (1906).—P. Taylor in Kew Bull. 18: 53 (1964); in Fl. Afr. Centr., Lentibulariaceae: 10 (1972); in F.T.E.A., Lentibulariaceae: 7 (1973). Type from S. Africa (Cape Prov.).
    Utricularia hians A.DC. in DC. Prodr. 8: 25 (1844). Type from Madagascar.
    Utricularia lingulata Baker in Journ. Linn. Soc., Bot. 20: 216 (1883). Type from Madagascar.
    Utricularia prehensilis var. *hians* (A.DC.) Kam. in Engl., Bot. Jahrb. 33: 103 (1902). Type as for *Utricularia hians*.
    Utricularia prehensilis var. *huillensis* Kam. in Engl., Bot. Jahrb. 33: 103 (1902) partly excl. syn. *Utricularia madagascariensis*. Type from Angola.
    Utricularia prehensilis var. *lingulata* (Baker) Kam. in Engl., Bot. Jahrb. 33: 103 (1902). Type as for *Utricularia lingulata*.
    Utricularia quadricarinata Suessenguth in Trans. Rhod. Sci. Ass. 43: 119 (1951). Type: Zimbabwe, Marondera, *Dehn* 32, 429 (M, syntypes).

Terrestrial herb. Rhizoids and stolons capillary, numerous, fasciculate from the base of the peduncle. Leaves numerous but usually decayed at anthesis, branch-opposed on the stolons, linear to narrowly oblanceolate, up to 10 cm. × 3 mm., 1–7-nerved. Traps scattered on the stolons and leaves, globose, shortly stalked, glandular, 0.6–1.5 mm. long; mouth basal, with 2 dorsal simple subulate appendages and a single, laterally compressed ventral appendage. Inflorescence erect or twining, 3–35 cm. high; peduncle filiform, glabrous; flowers 1–8, distant; scales few, similar to the bracts; bracts basifixed, ovate-deltoid, 1.5–2 mm. long, up to 5-nerved; bracteoles linear, shorter than the bract; pedicels erect, narrowly winged, as long as or longer than the fruiting calyx. Calyx lobes ovate, up to 5 mm. long at anthesis and 10 mm. long in fruit, more or less unequal, upper usually larger, with apex acute, obtuse or subacute, lower truncate or shortly bidentate. Corolla yellow, 8–20 mm. long; superior lip more or less twice as long as the upper calyx lobe, narrowly oblong to broadly spathulate with apex rounded, truncate or more or less emarginate; inferior lip more or less circular, with apex entire, bifid or obscurely 4-crenate; palate conspicuously raised, usually prominently longitudinally 4-ridged; spur subulate, acute, more or less as long as the inferior lip. Filaments linear, curved; anther-theacae subdistinct. Ovary ovoid, dorsiventrally compressed; style short; stigma inferior lip rounded, superior lip smaller. Capsule ovoid, up to 5 mm. long, dehiscing by a longitudinal ventral slit with thickened margins. Seeds numerous, ovoid, 0.6–0.8 mm. long, usually verrucose; testa rather loose and corky, the cells distinct, elongate.

Zambia. N: Chinsali Distr., Mansha R., 7 km. W. of Shiwa Ngandu, 1440 m., fl. 28.v.1972, *Kornas* 1891 (K). W: Mwinilunga Distr., 0.8 km. S. of Matonchi Farm, fl. 11.xi.1938, *Milne-Redhead* 4531 (K). **Zimbabwe**. C: Harare, Ruwa R., 1966 m.,fl. 7.iv.1947, *Wild* 1961 (K; SRGH). E: Inyanga, Gairesi Ranch on Mozambique border, c. 10 km.N. of Troutbeck, 1933 m., fl. 17.xi.1956, *Robinson* 1924 (K; SRGH). **Malawi**. N: Nyika Plateau, Chelinda Bridge, Chelinda R., Nyika Game Reserve, 2250 m., fl. 21.xi.1967, *Richards* 22659 (K). C: Dedza, Ngoma, Chongoni Forest, fl. 16.xi.1970, *Salubeni* 1461 (K; SRGH). **Mozambique**. M: Maputo, 20 m., fl. 31.i.1979, *Schäfer* 6677 (K; LIJSC).
    Widespread in eastern and southern Africa from Ethiopia to Angola and S. Africa (eastern Cape Prov.) and in Madagascar. Bogs, marshes and seasonally flooded ground by lakes and rivers; from sea level (in the south) to 2100 m.

15. **Utricularia spiralis** Sm. in Rees, Cyclop. 37, no. 5 (1819).—Stapf in Dyer, F.T.A. 4: 482 (1906) partly.—Hutch. & Dalz., F.W.T.A. 2: 234 (1931) partly. Type from Sierra Leone.

Utricularia baumii Kam. in Engl., Bot. Jahrb. **33**: 102 (1902).—Stapf in Dyer, F.T.A. 4: 480 (1906).—Hutch., Botanist in S. Afr.: 528 (1946). Type from Angola.
Utricularia baumii var. *leptocheilos* Pellegrin in Bull. Soc. Bot. Fr. **61**: 16 (1914). Syntypes from Guineé & Ivory Coast.
Utricularia paradoxa Lloyd & G. Taylor in Contr. Gray Herb. **165**: 83 (1947) partly, inflorescence only. Type from Angola.
Utricularia spiralis var. *spiralis* P. Taylor in F.W.T.A., ed. 2, **2**: 378 (1963); in Kew Bull. **18**: 60 (1964); in Fl. Afr. Centr., Lentibulariaceae: 12 (1972); in F.T.E.A., Lentibulariaceae: 8 (1973). Type as for *Utricularia spiralis*.

Terrestrial herb. Rhizoids and stolons capillary, numerous from the base of the peduncle. Leaves few, usually decayed at anthesis, leaf-opposed on the stolons, linear, up to 5 cm. × 2.5 mm., 1—3-nerved. Traps numerous, globose, shortly stalked, glandular, 0.6—1.0 mm. long; mouth basal, with 2 dorsal simple subulate appendages. Inflorescence erect or usually twining, 20—70 cm. high; peduncle filiform, glabrous; flowers 3—10, usually distant; scales few, similar to the bracts; bracts basifixed, broadly ovate, 2 mm. long; bracteoles linear-lanceolate, usually shorter than the bracts; pedicels erect or ascending, 5—15 mm. long, flattened or more or less narrowly winged, always longer than the fruiting calyx. Calyx lobes subequal 3—6 mm. long, ovate to narrowly ovate with apex of upper acute, of lower minutely bidentate. Corolla usually violet with a dark blue, greenish, yellow or white spot in the throat, rarely wholly yellow or white, 1—3 cm. long; superior lip oblong to circular; inferior lip circular; palate raised; spur subulate, usually curved, acute. Filaments linear; anther-thecae subdistinct; ovary ovoid; style indistinct; stigma-lips short, truncate, subequal. Capsule narrowly ovoid, dehiscing by longitudinal dorsal and ventral slits; capsule-wall of uniform thickness. Seeds numerous, globose, 0.2—0.3 mm. in diam.; testa thin, cells distinct, more or less isodiametric.

Zambia. N: Kasama Distr., Mungwi, fl. 20.vi.1960, *Robinson* 3754 (K). W: Mwinilunga Distr., 20 km. from Mwinilunga along Rd. to Kalene Hill, fl. 21.xi.1972, *Strid* 2559 (K).
Widespread but somewhat scattered in Africa from Guineé to Angola. In peaty or sandy soil in swamps and marshes; sea level to 1860 m.

16. Utricularia tortilis Oliver in Journ. Linn. Soc., Bot. **9**: 150.—Stapf in Dyer, F.T.A. 4: 483 (1906).—R.E. Fries, Wiss. Ergebn. Schwed. Rhod.-Kongo-Exped. **1**: 298 (1916). TAB. 2. Type from Angola.
Utricularia spiralis sensu Stapf in Dyer, F.T.A. 4: 482 (1906) partly.—Hutch. & Dalz., F.W.T.A. **2**: 234 (1931) partly non Sm.
Utricularia falcata Good in Journ. Bot. **62**: 162 (1924). Type from Angola.
Utricularia gyrans Suessenguth in Trans. Rhod. Sci. Ass. **43**: 119 (1951). Type: Zimbabwe, Marondera, *Dehn* 33 (M, holotype).
Utricularia spiralis var. *tortilis* (Oliver) P. Taylor in Taxon **12**: 294 (1963); in F.W.T.A., ed 2, **2**: 378 (1963); in Kew Bull. **18**: 62 (1964); in Fl. Afr. Centr., Lentibulariaceae: 12 (1972); in F.T.E.A., Lentibulariaceae: 8 (1973). Type as for *Utricularia tortilis*.

Terrestrial herb. Rhizoids and stolons capillary, numerous from the base of the peduncle. Leaves few, usually decayed at anthesis, leaf-opposed on the stolons, linear, up to 2 cm. × 0.5—15 mm., 1—3 nerved. Traps numerous, globose, shortly stalked, glandular, 0.6—1.0 mm. long; mouth basal, with 2 dorsal simple subulate appendages. Inflorescence erect or usually twining, 5—40 cm. high; peduncle filiform, glabrous; flowers 1—6, usually distant; scales few, similar to the bracts; bracts basifixed, broadly ovate, more or less 2 mm. long; bracteoles linear-lanceolate, usually shorter than the bracts; pedicels erect or ascending, 2—15 mm. long, flattened or more or less narrowly winged, usually longer than the fruiting calyx. Calyx-lobes subequal, 3—6 mm. long, with apex of upper acute, of lower minutely bidentate. Corolla usually violet with a dark blue, greenish, yellow or white spot in the throat, rarely wholly yellow or white, 5—10 mm. long; superior lip narrowly oblong; inferior lip circular; palate raised; spur subulate, usually curved, acute or rarely obtuse. Filaments linear; anther-thecae subdistinct; ovary ovoid; style indistinct; stigma-lips short, truncate, subequal. Capsule narrowly ovoid, dehiscing by longitudinal ventral and dorsal slits, capsule-wall of uniform thickness. Seeds numerous, globose, 0.3—0.35 mm. in diam.; testa thin, cells distinct, more or less isodiametric.

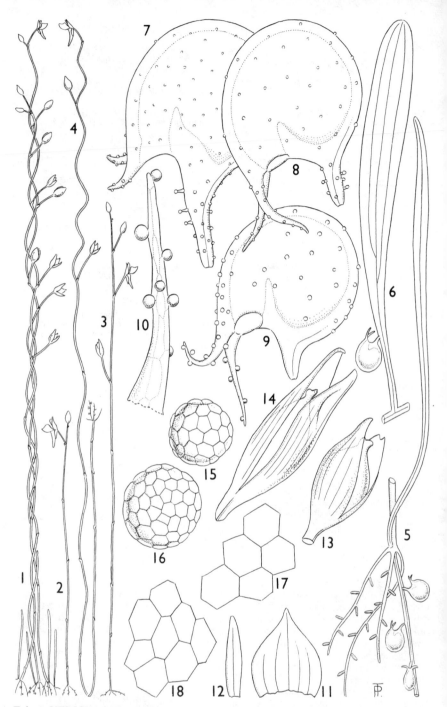

Tab. 2. UTRICULARIA TORTILIS. 1, group of flowering and fruiting plants, twining (×1), *Watermeyer* 163; 2 & 3, flowering and fruiting peduncles, not twining (×1), 2–3 from *Melville & Hooker* 282; 4, flowering and fruiting peduncle, twining (×1), *Milne-Redhead & Taylor* 9830; 5, base of plant showing stolons, traps, rhizoids, leaf and peduncle base (×8), *Fanshawe* 3194; 6, leaf (×8), *Baldwin* 13005; 7, traps, lateral view (×60), *Melville & Hooker* 282; 8, traps, lateral view (×60), *Milne-Redhead & Taylor* 10732; 9, traps, lateral view (×60); 10, appendage from trap mouth (×150), 9–10 from *Milne-Redhead & Taylor* 9830; 11, bract (×12); 12, bracteole (×12), 11–12 from *Germain* 2051; 13, capsule, lateral view (×8), *Melville & Hooker* 282; 14, capsule, lateral view (×8); 15, seed (×60), 14–15 from *Fanshawe* 3194; 16, seed (×60), *Meikle* 1092; 17, testa cells (×150), *Baldwin* 13005; 18, testa cells (×150), *Milne-Redhead & Taylor* 10732.

Botswana. N: Okavango Swamps, c. 10 km. S. of Seronga, fl. 29.ix.1954, *Story* 4796 (K; PRE). **Zambia.** B: Kalabo, c. 5 km. S. of Kalabo, fl. 16.xi.1959, *Drummond & Cookson* 6536 (K; SRGH). N: Mbala Distr., Kanbole Escarpment near Ngozi R., 1500 m., fl. 6.vi.1957, *Richards* 10044 (K). W: Kitwe, fl. 20.iv.1957, *Fanshawe* 3194 (K). S: c. 2.5 km. E. of Choma, 1433 m., fl. 27.iii.1955, *Robinson* 1186 (K; SRGH). **Zimbabwe.** W: Matobo, Farm Besna Kobila, 1600 m., fl. vi.1957, *Miller* 4413 (K; SRGH). C: Harare, Chinamora Reserve, Ngoma kurira, 1835 m., fl. 25.iii.1952, *Wild* 3782 (K; SRGH).

Widespread in Africa from Senegal to Angola. In peaty or sandy soil in swamps and marshes; sea level to 1860 m.

There is some evidence of hybridization between this and *Utricularia spiralis*.

17. **Utricularia baoulensis** A. Chev. in Mém. Soc. Bot. Fr. 8: 186 (1912).—P. Taylor in F.W.T.A., ed. 2, 2: 378 (1963); in Kew Bull. 18: 69 (1964); in Fl. Afr. Centr., Lentibulariaceae: 6 (1972); in F.T.E.A., Lentibulariaceae: 8 (1973). Type from Ivory Coast.
    Utricularia scandens Oliver in Journ. Linn. Soc., Bot. 3: 181 (1859) non Benj. (1847).
    Utricularia spiralis sensu Hutch. & Dalz., F.W.T.A. 2: 234 (1931) partly non Sm.

Terrestrial herb. Rhizoids and stolons capillary, few from the base of the peduncle. Leaves few, scattered on the stolons, linear, up to 3 cm. × 0.4—1.0 mm. 1-nerved. Traps few on the stolons and leaves, globose, 0.8—1.2 mm. long, shortly stalked, glandular; mouth basal, with 2 dorsal subulate branched appendages. Inflorescence twining up to 20 cm. high; Peduncle filiform, glabrous; flowers 2—5, distant; scales few, similar to the bracts; bracts basifixed, ovate, more or less 1.2 mm. long; bracteoles linear-lanceolate, more or less half as long as the bract; pedicels erect at anthesis, strongly decurved in fruit, more or less as long as the calyx, flattened and narrowly winged. Calyx lobes ovate, subequal, more or less 2 mm. long at anthesis, 3.5—4 mm. long, with the lower lobe slightly relatively longer in fruit. Corolla pale blue or mauve, 3—4 mm. long; superior lip oblong with apex truncate, slightly longer than the upper calyx lobe; inferior lip circular, with apex obscurely 3-crenate; palate scarcely raised; spur narrowly conical, obtuse. Filaments linear; anther-thecae subdistinct. Ovary ovoid; style short, distinct; stigma inferior lip circular, superior lip very short, truncate. Capsule broadly ovoid, dehiscing by a single longitudinal ventral slit with thickened margins in the upper half. Seeds numerous, ovoid to ellipsoid, more or less 0.3 mm. long; testa loose, corky, the cells distinct, elongate.

Zambia. W: 7 km. E. of Chizela, fl. 27.iii.1961, *Drummond & Rutherford-Smith* 7445 (K; SRGH) S: 4.8 km. E. of Mapanza, 1166 m., fl. 28.iii.1954, *Robinson* 643 (K). **Mozambique.** Z: Quelimane, Namagoa, fl. 18.v.1954, *Faulkner* K269a (K).

Scattered through tropical Africa and in Madagascar, India, China (Hairan), Thailand, Java, Phillipines and Australia (Queensland). Marshes and swamps; sea level to 1000 m.

18. **Utricularia striatula** Sm. in Rees, Cyclop. 37, no.17 (1819).—Stapf in Dyer, F.T.A. 4: 486 (1906).—P. Taylor in F.W.T.A., ed. 2, 2: 378 (1963); in Kew Bull. 18: 91 (1964); in Fl. Afr. Centr., Lentibulariaceae: 32 (1972); in F. T.E.A., Lentibulariaceae: 13 (1973). Type from Sierra Leone.
    Utricularia philetas Good in Journ. Bot. 62: 163 (1924). Type from Angola.

Epiphytic herb. Rhizoids and stolons capillary, numerous from the base of the peduncle. Leaves numerous and conspicuous at anthesis, rosulate at the peduncle base and scattered on the stolons, obovate to reniform, petiolate, 3—10 mm. 1—6 mm. Traps numerous, globose or ovoid. 0.6—0.8 mm. long, stalked; mouth lateral, with a deeply bifid dorsal appendage, densely covered with glandular hairs. Inflorescence erect, straight, 1—15 cm. high; peduncle capillary, glabrous; flowers 1—10, distant; scales few, similar to the bracts; bracts medifixed, lanceolate, 1.5—2 mm. long; bracteoles similar to the bracts but smaller; pedicels capillary, 2—6 mm. long, spreading or sometimes deflexed in fruit. Calyx lobes very unequal and accrescent; upper lobe circular-obcordate, with apex emarginate, 1.5—2.5 mm. long at anthesis; lower lobe ovate-oblong, more or less half as long as the upper. Corolla white or mauve with a yellow spot on the throat, 3—6 mm. long or sometimes much reduced (cleistogamous); superior lip minute, much shorter than the upper calyx lobe, very shortly 2-lobed; inferior lip circular in outline, 3—10 mm. long, more or less regularly 5-lobed;

palate slightly raised; spur subulate, curved or straight, more or less as long as the inferior lip. Filaments filiform; anther-thecae subdistinct. Ovary globose, adnate to the base of the upper calyx lobe; style short; stigma inferior lip semi-circular, superior lip obsolete. Capsule globose, obliquely dorsiventrally compressed, carinate on the ventral surface, dehiscing by a longitudinal ventral slit. Seeds numerous, ovoid, 0.25 mm. long, covered with short glochidiate processes.

Zambia. N: Mbala, Mwenda Hills, fl. 5.v.1957, *Richards* 9586 (K).
From Guineé to Ethiopia and Angola and widespread in tropical Asia from India to New Guinea. Wet mossy rocks and tree trunks; sea level to 2250 m.

19. **Utricularia appendiculata** Bruce in Bull. Misc. Inf. Kew **1933**: 475 (1934).—P. Taylor in Kew Bull. **18**: 95 (1964); in Fl. Afr. Centr., Lentibulariaceae: 30 (1972); in F.T.E.A., Lentibulariaceae: 13 (1973). Type from Tanzania.

Terrestrial herb. Rhizoids and stolons capillary; numerous from the base of the peduncle. Leaves rosulate but usually decayed at anthesis, linear, thalloid, up to 3 cm. × 1.5−3 mm. 1-nerved. Traps numerous, ovoid, 0.8−1.0 mm. long, stalked; mouth lateral, oblique, with a single dorsal usually 3-branched, more or less recurved appendage. Inflorescence filiform, twining, up to 60 cm. high; peduncle glabrous above, minutely papillose at the base; flowers 1−8, distant; scales numerous, similar to the bracts; bracts medifixed, more or less 3 mm. long, narrowly lanceolate, acute at both extremities; bracteoles similar but somewhat smaller; pedicels erect, 2−3 mm. long. Calyx lobes very unequal; upper broadly rhomboid, more or less 3 mm. long; lower ovate or ovate-oblong, much narrower than the upper, with apex truncate. Corolla white, cream or pale yellow, 5−9 mm. long; superior lip oblong, scarcely exceeding the upper calyx lobe, with apex truncate or emarginate; inferior lip circular, with apex entire or emarginate; palate raised, obscurely 2-gibbous; spur narrowly cylindrical, curved, more or less as long as the inferior lip. Filaments filiform, twisted; anther thecae subdistinct. Ovary ovoid; style very short; stigma inferior lip semi-circular, superior much smaller, deltoid or oblong. Capsule globose, obliquely dorsiventrally compressed, dehiscing by a longitudinal ventral slit. Seeds very numerous, narrowly cylindrical; testa longitudinally striate and with numerous short rough conical papillae.

Zambia. N: Inono R. Source, 1666 m., fl. 28.ii.1955, *Richards* 4710 (K; SRGH). W: Mwinilunga Distr., fl. 21.xii.1937, *Milne-Redhead* 3758 (BR; K). **Zimbabwe**. E: Chimanimani Distr., Stonehenge Plateau, 1833 m., fl. 2.ii.1957, *Phipps* 421 (K; SRGH). **Malawi**. N: Nkhata Bay Distr., 36.8 km. SW. of Mzuzu on Viphya Plateau, Lwafwa Drift, 1833 m., fl. 15.v.1976, *Pawek* 11272 (K; MAL; MO). **Mozambique**. MS: Chimanimani Mts., c. 3.2 km. SE. of the Saddle between St. Georges and Poacher's Caves, c. 1666 m., fl. 12.iv.1967, *Grosvenor* 397 (K; SRGH).
Also in Cameroun, Katanga Gabon, Zaire, Rwanda, Burundi, Uganda, Tanzania amd Madagascar. Bogs and marshes; 200−1860 m.

20. **Utricularia subulata** L., Sp. Pl.: 18 (1753).—Stapf in Dyer, F.T.A. **4**: 485 (1906).—P. Taylor in F.W.T.A., ed. 2, **2**: 380 (1963); in Kew Bull. **18**: 81 (1964); in Fl. Afr. Centr., Lentibulariaceae: 34 (1972); in F.T.E.A., Lentibulariaceae: 14 (1973). TAB. 3. Type from U.S.A. (Virginia).
   Utricularia perpusilla A.DC. in DC. Prodr 8: 25 (1844). Type from Madagascar.
   Utricularia angolensis Kam. in Engl., Bot. Jahrb. **33**: 104 (1902). Type from Angola.
   Utricularia subulata var. *minuta* Kam. in Engl., Bot. Jahrb. **33**: 105 (1902). Type from Zaire.
   Utricularia triloba Good in Journ. Bot. **62**: 162 (1924). Type from Angola.
   Utricularia rendlei Lloyd in Journ. Bot. **73**: 42 (1935). Type: Zambia, Victoria Falls, Livingstone Is., *Llyod* 3 (BM, holotype).

Terrestrial herb. Rhizoids and stolons capillary, few from the base of the peduncle. Leaves linear, 1−2 cm. × more or less 0.5 mm., 1-nerved.Traps very numerous, globose or ovoid, 0.2−0.5 mm. long, stalked, the stalk slightly shorter than the trap; mouth lateral, oblique, with two dorsal branched appendages. Inflorescence erect, up to 25 cm. high; peduncle capillary, smooth above, minutely papillose below; inflorescence-axis usually flexuous or zigzag; floral

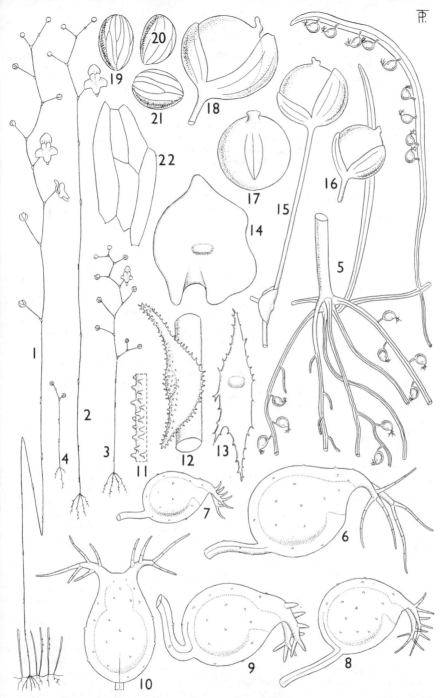

Tab. 3. UTRICULARIA SUBULATA. 1, flowering and fruiting plant (×1), *Adames* 93;
2, flowering peduncle (×1), *Milne-Redhead & Taylor* 9490; 3, flowering and fruiting
peduncle (×1), *Miller* 2254; 4, fruiting peduncle, cleistogamous (×1), *Jordan* 1066;
5, base of flowering plant, showing stolons, rhizoids, traps, leaves and peduncle base
(×8); 6 & 7 traps, lateral view (×60), 5–7 from *Milne-Redhead & Taylor* 9490a; 8,
traps, lateral view (×60), *Brenan & Greenway* 8074; 9, traps, lateral view (×60), *Ross*
1507; 10, trap, dorsal view (×60); 11, papillae on peduncle base (×150), 10–11 from
Milne-Redhead & Taylor 9490a; 12, scale on lower peduncle (×30), *Harley* 1851; 13,
scale, flattened (×30); 14, bract, flattened (×30), 13–14 from *Schmitz* 3791; 15,
capsule, with pedicel and bract (×12); 16, capsule, lateral view (×12); 17, capsule,
abaxial view (×12), 15–17 from *Milne-Redhead & Taylor* 7779b; 18, capsule, lateral
view (×12), *Ross* 1507; 19, seed (×60), *Meikle* 1095; 20, seed (×60), *Ross* 1507; 21,
seed (×60), *Wild* 3781; 22, testa cells (×150), *Melville & Hooker* 235.

internodes slightly longer than the pedicels; scales few, peltate, more or less as long as the bracts, narrowly elliptic, acuminate at both extremities, minutely papillose; bracts peltate, circular, 0.75−1.0 mm. long, smooth, clasping the base of the pedicel; bracteoles absent; pedicels capillary, ascending, 2−10 mm. long. Calyx lobes subequal, broadly ovate, more or less 1 mm. long at anthesis, slightly accrescent, obscurely 5-nerved, with apex obtuse or truncate. Corolla yellow, normally 6−10 mm. long, sometimes reduced (in cleistogamous flowers) to 2 mm.; superior lip broadly ovate, with apex rounded, 2−3 times as long as the upper calyx lobe; inferior lip circular in outline, deeply 3-lobed; palate raised, bigibbous; spur subulate, parallel with and more or less equalling the inferior lip or (in cleistogamous flowers) much reduced and saccate. Filaments filiform; anther-thecae confluent. Ovary globose; style very short; stigma inferior lip circular, superior lip more or less obsolete. Capsule globose, 1−1.5 mm. long, dehiscing by a small ovate ventral pore. Seeds numerous, ovoid, 0.2−0−24 mm. long, longitudinally striate.

Botswana. N: Gidiba Isl., fl. 14.ix.1976, *Smith* 1772 (K; SRGH). **Zambia.** B: Kalabo, 4.8 km. S. of Kalabo, fl. 16.xi.1959, *Drummond & Cookson* 6537 (K; SRGH). N: Mansa, fl. 8.x.1947, *Brenan & Greenway* 8074 (BR; EA; FHO; K; LISC; SRGH). W: Mwinilunga, fl. 21.xii.1937, *Milne-Redhead* 3759 (K). C: Serenje Distr., Mulembo R., on Mukuku Rd. 56 km. of Great North Rd., 1230 m., fl. 6.v.1972, *Kornas* 1705 (K). S: Livingstone, fl. 29.viii.1947, *Brenan & Greenway* 7773 (K; FHO). **Zimbabwe.** W: Matobo, Farm Besna Kobila, 1500 m., fl. iii.1954, *Miller* 2254 (K; MO; SRGH). C: Harare, Chinamora Reserve, Ngomakurira 1833 m., fl. 25.iii.1952, *Wild* 3781 (K; SRGH). E: Mutare, Nusa Plateau, fl. iii.1935, *Gilliland* Q1620 (BM; K). **Malawi.** N: Rumphi Distr., Nyika Plateau 12.8 km. N. of M1, fl. 11.iii.1978, *Pawek* 14073 (K; MO). C: Lilongwe Distr., Dzalanyama Forest Reserve., above Chaulongwe Falls, 1300 m., fl. 26.iv.1970, *Brummitt* 10173 (K). **Mozambique.** MS: Beira, Cheringoma Coast, Ngamazura, fl. v.1973, *Tinley* 2909 (K; SRGH). GI: Bilene, fl. 5.xii.1979, *Schäfer* 7107 (K; LMU).

Very widespread in Africa from Senegal to Sudan, Angola and S. Africa (Natal), also in Madagascar, N. & S. America and India to Australia. Wet peaty or sandy soil in grassland; sea level to 1800 m.

21. **Utricularia australis** R. Br., Prodr. Fl. Nov. Holl.: 430 (1810).—P. Taylor in Fl. Afr. Centr., Lentibulariaceae: 48 (1972); in F.T.E.A., Lentibulariaceae: 20 (1973). Type from Australia.

     Utricularia incerta Kam. in Engl., Bot. Jahrb. **33**: 111 (1902).—Stapf in Dyer, F.T.A. **4**: 496 (1906). Type from Sudan.

     Utricularia vulgaris sensu P. Taylor in Kew Bull. **18**: 171 (1964), non L.

Aquatic herb, perennating by winter buds (turions). Stolons 2 or 3 from the base of the peduncle, filiform, terete, glabrous, up to 50 cm. long or more, 0.5−1.5 mm. thick; internodes 3−10 mm; rhizoids few (2−4) from the peduncle base, capillary, 1−12 cm. long. Leaves very numerous, 2-branched from the base, each branch 1−5 cm. long, ovate to lanceolate in outline, pinnately branched; pinnae alternate, repeatedly dichotomously forked; ultimate segments capillary, setulose. Traps usually numerous, lateral on the leaf-segments just above the point of bifurcation, obliquely ovoid, stalked, 1−2 mm. long; mouth lateral, oblique, with 2 dorsal slender more or less branched hairs, and a variable number of ventral simple hairs. Inflorescence erect, up to 15 cm. high; peduncle usually straight at anthesis but becoming very flexuous, 1−2 mm. thick, smooth and glabrous; flowers 4−6, more or less congested at anthesis, inflorescence-axis elongating after anthesis; scales 1−2(3), a short distance below the lowermost flower, similar to the bracts; bracts basifixed, circular, more or less 3 mm. long, decurrent; bracteoles absent; pedicels filiform, 1−2 cm. long, erect at anthesis, elongating and spreading or deflexing after anthesis. Calyx lobes subequal, ovate, more or less 3 mm. long; superior lobe with apex rounded, hyaline; lower lobe truncate or emarginate. Corolla pale yellow, more or less 15 mm. long; superior lip circular or broadly ovate, truncate, 2−3 times as long as the upper calyx lobe; inferior lip oblate or reniform, more or less as long as the superior lip and up to twice as wide; palate raised, gibbous; spur stout, conical, slightly curved, obtuse, usually shorter lip and up to twice as wide; palate raised, gibbous; spur stout, conical, slightly curved, obtuse, usually shorter than the inferior lip, with shortly stalked glands on the whole of the inner surface of the distal half. Filaments filiform; anther-thecae confluent. Ovary globose, minutely lepidote; style distinct, almost as long as the ovary; stigma inferior lip semi-circular, ciliate, superior almost obsolete. Capsules very rarely produced.

Botswana. N: Linyanti R., at Shaile, 1000 m., fl. 28.x.1972, *Gibbs-Russell* 2401 (K; SRGH). **Zimbabwe.** W: Hwange, 40 km. NE. of Robins Camp in controlled Hunting Area near Game Fence, fl. 24.x.1968, *Rushworth* 1233 (K; SRGH).

Also in Zaire, Sudan, Uganda, Tanzania, S. Africa (Transvaal and Natal), temperate Eurasia to India, Malesia, Australia and New Zealand.

22. **Utricularia inflexa** Forsk., Fl. Aegypt.-Arab., Descr. Pl.: 9 (1775).—P. Taylor in Fl. Afr. Centr., Lentibulariaceae: 38 (1972); in F.T.E.A., Lentibulariaceae: 16 (1973). Type from Egypt.

Utricularia thonningii Schumach. in Schumach. & Thonn., Beskr. Guin. Pl.: 12 (1827).—Stapf in Dyer, F.T.A. 4: 487 (1906). Type from Guineé.

Utricularia stellaris var. *inflexa* (Forsk.) C.B Cl. in Hook. f., Fl. Brit. Ind. 4: 329 (1884).—Dur. & Schinz, Etud. Fl. Congo 1: 214 (1896). Type as for *Utricularia inflexa*.

Utricularia inflexa var. *remota* Kam. in Deutsch. Bot. Gesellsch. 12: 5 (1894). Type from Zanzibar.

Utricularia inflexa var. *tenuifolia* Kam. in loc. cit. Type from Madagascar.

Utricularia oliveri Kam. in Deutsch. Bot. Gesellsch. 12: 4 (1894). Syntypes from Senegal, Sudan & Zanzibar.

Utricularia oliveri var. *fimbriata* Kam. in tom. cit. 4. Type from Cameroun.

Utricularia oliveri var. *schweinfurthii* Kam. in loc. cit. Type from Sudan.

Utricularia thonningii var. *major* Kam. in Engl. Bot. Jahrb. 33: 109 (1902). Syntypes from Egypt and Sudan.

Utricularia thonningii var. *laciniata* Stapf in Dyer, F.T.A. 4: 488 (1906). Type from Zanzibar.

Utricularia stellaris sensu Perrier in Mém. Inst. Sci. Madag., Sér. B, 5: 188 (1955), partly, non L. f.

Utricularia inflexa var. *inflexa* P. Taylor in Mitt. Bot. Staats. München 4: 95 (1961); in F.W.T.A., ed. 2: 380 (1963); in Kew Bull. 18: 156 (1964). Type as for *Utricularia inflexa*.

Aquatic herb. Stolons up to 1 mm. long or more, terete, smooth and glabrous; internodes 3−10(15) mm. Leaves very numerous, digitately divided into 3−6 primary segments and usually auricled at the base; primary segments 2−6 cm. long, more or less lanceolate in outline, pinnately divided, the rhachis filiform or more or less inflated and up to 2 mm. thick; pinnae repeatedly dichotomously forked; ultimate segments capillary, minutely setulose; auricles when present reniform or circular, up to 10 mm. long, margin usually denticulate or more or less divided into setulose capillary segments. Floats verticillate or subverticillate on the lower half of the peduncle, 3−10, inflated, narrowly cylindrical to fusiform, 2−5 cm. long, bearing a variable number of reduced or sometimes well developed leaf segments at the apex. Traps usually numerous, lateral near the base of the ultimate and penultimate leaf segments, broadly ovoid, shortly stalked, 1−3 mm. long; mouth lateral, naked or with 2 dorsal short simple or sparsely branched hairs, and sometimes with a few shorter ventral simple hairs. Inflorescence erect, lateral, 3−20 cm. high; flowers 2−15, congested at anthesis; inflorescence-axis elongating in fruit; peduncle filiform to relatively stout, straight, smooth and glabrous; scales absent; bracts basifixed, ovate, 1−2 mm. long; bracteoles absent; pedicels filiform, erect or spreading at anthesis, 1.5−5 mm. long, strongly reflexed and thickened in fruit. Calyx lobes subequal, broadly ovate, with apex rounded or lower sometimes emarginate, more or less 3 mm. long at anthesis, strongly acrescent, becoming fleshy, up to 10 mm. long and enclosing the capsule. Corolla white or yellow, usually variously marked with red or purple lines, 7−10 mm. long, more or less densely covered externally with glandular hairs; superior lip broadly ovate, 1.5−2 times as long as the upper calyx lobe, with apex truncate or emarginate; inferior lip circular, with apex rounded or emarginate; palate raised, bigibbous; spur cylindrical, more or less as long as the inferior lip; filaments linear, dilated above; anther-theacae subdistinct. Ovary globose, minutely squamulate; style distinct, as long as the ovary; stigma inferior lip circular, superior obsolete. Capsule globose, circumscissile, more or less 5 mm. in diam., with the style often considerably elongated. Seeds numerous, prismatic, 4−6 angled, more or less as wide as high, more or less winged on the angles, hilum prominent; testa-cells distinct, elongated, more or less 0.05 mm. long.

Zambia. N: Mbala Distr., Tyendwe Valley, Lufubu R., 780 m., fl. 16.viii.1964, *Richards* 19081 (K). S: Mazabuka, in the Kafue R., fl. 4.v.1964, *van Rensburg* 2918 (K). **Zimbabwe.** N: Binga inlet, Lake Kariba, fl. 7.iv.1960, *Phipps* 2800 (K; SRGH). **Malawi.** N: Mzimba Distr., Viphya Plateau, 2000 m., fl. 1948, *Benson* 1445 (BM). S: Zomba, Lake Chilwa, near boat landing, fl. 20.vii.1972, *Gibbs-Russell* 2121 (K; SRGH). **Mozambique.** MS: Beira Distr., Dondo Area, 3 km. SE. of Pungwe Bridge, 12−13 km., fl. 10.vii.1972, collector ?(K; Durban College).

Very widespread in Africa from Mauritania to Egypt and Namibia and S. Africa (Natal), also in Madagascar and India. Still or slow-flowing shallow or deep water, in rivers, lakes and marshes; sea level to 1700 m.

23. **Utricularia stellaris** L. f., Suppl.: 86 (1781).—Stapf in Dyer, F.T.A. 4: 489 (1906).— Hutch. & Dalz., F.W.T.A. 2: 234 (1931).—P. Taylor in Fl. Afr. Centr., Lentibulariaceae: 42 (1972); in F.T.E.A., Lentibulariaceae: 16 (1973). Type from India.

Utricularia stellaris var. *dilatata* Kam. in Deutsch. Bot. Gesellsch. 12: 3 (1894). Syntypes from Tanzania and Madagascar.

Utricularia flexuosa var. *parviflora* Kam. in Engl., Bot. Jahrb. 33: 108 (1902). Type from Zaire.

Utricularia stellaris var. *breviscapa* Kam. in Engl., in loc. cit. Syntypes from Namibia and S. Africa (Cape Prov.).

Utricularia stellaris var. *filiformis* Kam. in Engl., in loc. cit. Syntypes from Senegal, Sudan and S. Africa (Natal).

Utricularia trichoschiza Stapf in Dyer, F.T.A. 4: 488 (1906). Syntypes from Nigeria.

Utricularia inflexa var. *stellaris* (L. f.), P. Taylor in Mitt. Bot. Staatss. München 4: 96 (1961); in F.W.T.A., ed 2, 2: 380 (1963); in Kew Bull. 18: 189 (1964) partly excl. syn. *Utricularia muelleri* Kam.

Aquatic herb. Stolons up to 1 m. × 2.5 mm., terete, smooth and glabrous; internodes 5−20 mm. Leaves very numerous, digitately divided into 3−6 primary segments and usually auricled at the base; primary segments 1−6 cm. long, pinnately divided; rhachis filiform or more or less inflated, up to 2 mm. thick; pinnae repeatedly dichotomously forked; ultimate segments capillary, minutely setulose; auricles more or less deeply divided into ciliate filiform segments. Floats verticillate or subverticillate on the upper half of the peduncle, 5−7, inflated, ellipsoid, 0.5−2 cm. long, bearing a few reduced leaf-segments at the apex. Traps usually numerous, lateral near the base of the ultimate and penultimate leaf segments, broadly ovoid, shortly stalked, 1−3 mm. long; mouth lateral, naked or with 2 dorsal short simple or sparsely branched hairs and with a few ventral short simple hairs. Inflorescence erect, lateral, 3−30 cm. high; flowers 2−16, more or less congested at anthesis; inflorescence-axis elongating with maturity; filiform, straight, smooth and glabrous; scales absent; bracts basifixed, broadly ovate, 2−3 mm. long; bracteoles absent; pedicels filiform, erect or spreading at anthesis, 1.5−5 mm. long, deflexed or decurved and more or less elongating and thickening in fruit. Calyx-lobes subequal, broadly ovate, decurrent, more or less 3 mm. long at anthesis, up to 5 mm. and reflexed in fruit; upper lobe with apex rounded; lower lobe with apex truncate or emarginate. Corolla yellow, usually with reddish lines on the palate, 7−10 mm. long, more or less densely covered externally with glandular hairs; superior lip broadly ovate, 1.5−2 times as long as the upper calyx lobe, with apex truncate or emarginate; inferior lip circular or oblate, truncate, emarginate or 3-crenate; palate raised, bigibbous; spur cylindrical, more or less as long as the inferior lip. Filaments linear, dilated above; anther-thecae subdistinct. Ovary globose; style short but distinct; stigma inferior lip semi-circular, ciliolate, superior very short or obsolete. Capsule globose, circumscissile, more or less 5 mm. in diam. with the style more or less elongated. Seeds numerous, prismatic, 4−6 angled, more or less 3 times as wide as high, usually narrowly winged on the angles; testa cells distinct, elongated more or less 0.1 mm. long.

Botswana. N: Toromoja, Botletle R., 1000 m., fl. 22.iv.1971, *Thornton* 7 (K; SRGH). **Zambia.** B: Mongu, flood plain, fl. 30.iii.1966, *Robinson* 6899 (K). N: Mporokoso Distr., Mweru- wa- Ntipa, 1050 m., fl. 6.iv.1957, *Richards* 9067 (K; SRGH). W: Kasempa Distr., 7 km. E. of Chizela, fl. 27.iii.1961, *Drummond & Rutherford-Smith* 7424 (K; SRGH).S: 4.8 km., SW. of Mapanza, 1166 m., fl. 14.iii.1954, *Robinson* 615 (K; SRGH). **Zimbabwe.** W: Hwange National Park, small Mabwa Pan 14.4 km., S. of Main Camp, 1000 m., fl. 17.iv.1972, *Gibbs-Russell* 1605 (K; SRGH). C: Marondera, Chikokorana Pans, Chiota T.T.L., 1200 m., fl. 29.iv.1972, *Gibbs-Russell* 1988 (K; SRGH). E: Chimanimani, Tarka

Dam, fl. iv.1969, *Goldsmith* 32/69 (K; SRGH). S: Mwenezi, Pan W. of Nuanetsi R., 4.8 km. downstream from Malapate, fl. 25.iv.1961, *Drummond & Rutherford-Smith* 7523 (K; SRGH). **Malawi**. C: Kasungu Game Reserve, 1020 m., fl. 22.vi.1970, *Brummitt* 11639 (K). S: Chikwawa Distr., Lengwe Game Reserve, 100 m., fl. 6.iii.1970, *Brummitt* 8909 (K). **Mozambique**. Z: Quelimane Distr., Namagoa, fl. 23.vi.1946, *Faulkner* K266 (K; S). T: Between Tete and the sea coast, fl. vi & v.1860, *Kirk* 19269 (K). MS: Beira Distr., Cheringoma Section, Inhamissembe R., 8−9 m., fl. 14.vii.1972, *Ward* 7946 (K; Durban College). M: Inhaca Isl. 36.8 km., E. of Maputo 200 m., 26.ix.1957, *Mogg* 27547 (K).

Widespread in Africa from Mauritania and Egypt to S. Africa (Cape Prov.), also in Madagascar, Mauritius and Comoro Is., India to Vietnam, absent from Malesia, but in northern Australia. Still and slow flowing water in lakes, rivers and marshes; from sea level to 1700 m.

24. **Utricularia reflexa** Oliver in Journ. Linn. Soc., Bot. 9: 146 (1865).−Stapf in Dyer, F.T.A. 4: 492 (1906).−P. Taylor in F.W.T.A., ed. 2, 2: 380 (1963) excl. syn. *Utricularia kalmaloensis*. Syntypes from Nigeria and Angola.

Utricularia diploglossa Oliver in Journ. Linn. Soc., Bot. 9: 147 (1865).−Stapf in Dyer, F.T.A. 4: 434 (1906). Type from Nigeria.

Utricularia charoidea Stapf in Dyer, F.T.A. 4: 493 (1906). Type from Nigeria.

Utricularia pilifera A. Chev. in Mém. Soc. Bot. Fr. 8: 187 (1912). Type from Ivory Coast.

Utricularia bangweolensis R.E. Fries, Wiss. Ergebn. Schwed. Rhod.-Kongo-Exped. 1: 300 (1916). Type: Zambia, Lake Bangweulu, *Fries* 977 (UPS, holotype; K, Z,isotypes).

Utricularia magnavesica Good in Journ. Bot. 62: 165 (1924). Type from Angola.

Utricularia grandivesiculosa Czech in Mitt. Bot. Staatss. München 1: 344 (1952). Syntypes from Namibia.

Utricularia imerinensis Perrier in Mém. Inst. Sci. Madag., Sér. B, 5: 190 (1955). Type from Madagascar.

Utricularia reflexa var. *reflexa* P. Taylor in Kew Bull. 18: 164 (1964) excl. syn *Utricularia kalmaloensis*; in Fl. Afr. Centr., Lentibulariaceae: 45 (1972); in F.T.E.A., Lentibulariaceae: 9 (1973). Type as for *Utricularia reflexa*.

Utricularia reflexa var. *parviflora* P. Taylor in Kew Bull. 18: 168 (1964); in F.T.E.A., Lentibulariaceae: 19 (1973). Type from Tanzania.

Aquatic herb. Stolons filiform to relatively thick and fleshy, up to 50 cm. × 0.3−3.0 mm., or more or less densely covered with simple stellate hairs which extend also to the traps, basal part of the leaves and peduncle; internodes 2−10(15) mm. Rhizoids usually absent but when present 2−3 from the peduncle base, capillary, 3−5 mm. long. Leaves very numerous, digitately divided to the base into 2−5 primary segments, 3−1.30 mm. long, repeatedly dichotomously divided, the ultimate segments capillary, setulose. Traps inserted in the angles between the primary segments and between subsequent dichotomies, very variable in number and size, ovoid, shortly stalked, 1−6 mm. long; mouth lateral with 2 dorsal simple to conspicuously branched hairs and sometimes with ventral shorter simple hairs. Inflorescence lateral, 1.5−18 cm. high; flowers 1−3(4), distant, the lowermost sometimes very near the base of the peduncle; peduncle erect or rarely twining, filiform; scales absent; bracts basifixed, quadrate or circular, more or less encircling the base of the pedicel, 1−3 mm. long; bracteoles absent; pedicels filiform, erect at anthesis, usually strongly deflexed or decurved in fruit, 2−35 mm. long. Calyx lobes subequal, broadly ovate, 1−2 mm. long, scarcely accrescent, with apex obtuse or rounded. Corolla pale to quite deep yellow, usually with brown or reddish nerves, more or less densely covered externally with fine short hairs, rarely glabrous, 3−15 mm. long; superior lip broadly ovate or circular, with apex truncate or rounded, entire or emarginate, 1.5−3 times as long as the upper calyx lobe; inferior lip circular to subreniform, with apex more or less emarginate or 2-lobed; palate raised, bigibbous; spur cylindrical, more or less as long as the inferior lip. Filaments linear; anther-thecae more or less confluent. Ovary globose; style short or obsolete; stigma inferior lip semi-circular, sometimes ciliate, superior very short or obsolete. Capsule globose, circumscissile, 3−4 mm. in diam., few-many-seeded. Seeds lenticular, angular, 0.4−0.8 mm. wide, often narrowly winged on the angles; testa reticulate, cells distinct, isodiametric or more or less elongate.

Botswana. N: Savuti Marsh, 1000 m., 25.x.1972, *Gibbs-Russell* 2344 (K; SRGH). **Zambia**.

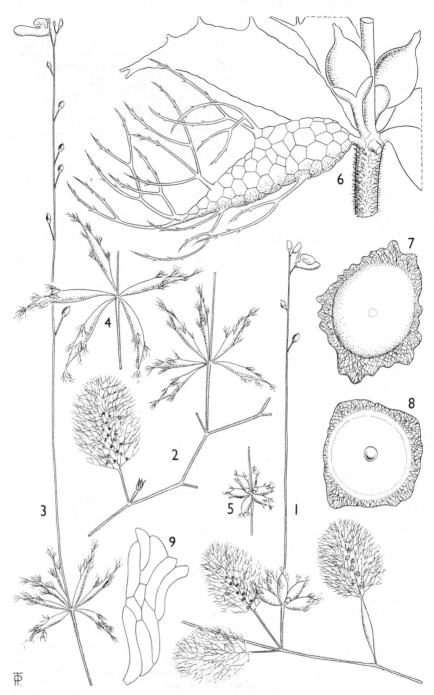

Tab. 4. UTRICULARIA BENJAMINIANA. 1, part of plant with leaves, with and without inflated stalk and inflorescence with float-leaves, cleistogamous and chasmogamous flowers and fruit (×1), *Robinson* 3796; 2, part of plant with a single-flowered cleistogamous inflorescence and base of chasmogamous inflorescence (×1), *Dinklage* 3277; 3, flowering and fruiting peduncle (×1), *Melville & Hooker* 457; 4, float-leaves (×1), *Deighton* 5973; 5, float-leaves with cleistogamous flowers (×1); 6, whorl of float-leaves with fruits from cleistogamous flowers (×8), 5—6 from *Matton* 2; 7, seed, apical view (×30); 8, seed, basal view (×30); 9, testa cells (×150), 7—9 from *Deighton* 5973.

B: Mongu, 30.iii.1966, *Robinson* 6898 (K). N: Mbala, Uningi Pans, 1833 m., fl. 10.iv.1964, *Vesey-Fitzgerald* s.n. (K; IRLCS). W: Chizela, 11.vi.1953, *Fanshawe* 77 (EA; K; SRGH). S: 8 km. E. of Choma, 1433 m., fl. 27.iii.1955, *Robinson* 1190 (K). **Zimbabwe**. W: Matobo, Farm Besna Kobila, 1600 m., fl. v.1955, *Miller* 2876 (K; SRGH). C: Marondera, fl. 26.ii.1948, *Corby* 22 (K; SRGH). E: Mutare, Nusa Plateau, iii.1935, *Gilliland* K1610 (BM; K).

Widespread in Africa from Senegal to Sudan and S. Africa (Natal), also in Madagascar. In deep or shallow still or flowing water, often entangled with other aquatic plants; from sea level to 2100 m.

25. **Utricularia benjaminiana** Oliver in Journ. Linn. Soc., Bot. 4: 176 (1860).—P. Taylor in F.W.T.A., ed. 2, 2: 380 (1963); in Kew Bull. 18: 179 (1964); in Fl. Afr. Centr., Lentibulariaceae: 36 (1972); in F.T.E.A., Lentibulariaceae: 15 (1973). TAB. 4. Type from Surinam.

Utricularia gilletii De Wild. & Dur. in Compt. Rend. Soc. Bot. Belg. 38: 40 (1900). Type from Zaire.

Utricularia villosula Stapf in Dyer, F.T.A. 4: 490 (1906). Type from Angola.

Utricularia paradoxa Lloyd & G. Taylor in Contr. Gray Herb. 165: 83 (1947) partly, as to vegetative parts only. Type from Angola.

Utricularia cevicornuta Perrier in Mém. Inst. Sci. Madag. Sér. B, 5: 190 (1955). Type from Madagascar.

Aquatic herb. Stolons filiform, up to 50 cm. long or more, flexuous, more or less densely villous; internodes 1−10 cm.; rhizoids absent. Leaves with 1 or 2 primary segments 1.5−4 cm. long, ovate in outline, pinnately divided; rhachis filiform throughout or more or less inflated, especially in the lower half, villous; pinnae alternate, the lowermost up to 1.5 mm. from the base of the rhachis, repeatedly dichotomously forked, ultimate segments capillary, sparsely setulose. Floats terminal on short lateral slightly inflated villous branches, verticillate, 4(5−6)−9, fusiform or narrowly cylindrical to ovoid or ellipsoid, 0.3−2.5 cm. long, bearing in the distal half a number of leaf-segments. Traps usually not numerous and often absent from some leaves, inserted in the angles between the leaf-rhachis and pinnae and between subsequent dichotomies of the leaves, ovoid, long-stalked, villous, 1−3 mm. long; mouth lateral, with 2 dorsal simple or more or less branched hairs and sometimes with one or more ventral shorter simple hairs. Inflorescence erect, 3−25 cm. high, terminal on the short stolon-branches, arising from the centre of the whorl of floats; flowers 2−10, the uppermost congested, those below more or less distant; cleistogamous flowers often present at the peduncle base among the floats; peduncle filiform, straight or flexuous, smooth and glabrous; scales absent; bracts basifixed, deltoid or ovate, more or less 1.5 mm. long; bracteoles absent; pedicels capillary, erect, 2−3 mm. long at anthesis, up to 10 mm. long in fruit. Calyx lobes subequal, circular, membranous, more or less 1 mm. long at anthesis, scarcely accrescent. Corolla mauve or pale puple with a yellow blotch on the palate or wholly white, 10−15 mm. long; superior lip oblong, 3−4 times as long as the upper calyx lobe, more or less auriculate at the base, deeply divided into 2 parallel lobes; inferior lip reniform, 7−10 mm. wide, with apex more or less emarginate; palate with a narrow raised more or less crenulate rim; spur cylindrical or botuliform, 2−3 times as long as the inferior lip, 8−10 mm. long, 2.5−3.5 mm. thick; corolla of cleistogamous flowers obsolete or nearly so. Filaments filiform; anther-thecae confluent. Ovary ovoid; ovules 3−20; style short but distinct; stigma inferior lip more or less quadrate, superior minute, deltoid or more or less obsolete. Capsule ellipsoid, 2.5−3.5 mm. long, circumscissile. Seeds few, sometimes only 2, lenticular with a narrow irregular wing, more or less 1 mm. in total diam.; testa-cells indistinct, elongated.

Botswana. N: Mojeje Dindinga Footpath, fl. 28.iii.1975, *Smith* 1320 (K; SRGH). **Zambia**. B: Mongu, 30.iii.1966, *Robinson* 6900 (K). N: Kasama Distr., Mungwi, 4.ix.1960, *Robinson* 3796 (K; SRGH). W: Chingola, fl. 19.iv.1954, *Fanshawe* 1125 (K).

Very widespread in Africa from Senegal to Namibia and S. Africa (Natal), in Madagascar and in Trinidad, Honduras, Venzuela, Guyana, Surinam, French Guiana and Brazil. In shallow to deep fresh or brackish water; from sea level to 1200 m.

26. **Utricularia foliosa** L., Sp. Pl.: 18 (1753).—Stapf in F.T.A. 4: 491 (1906).—P. Taylor in F.W.T.A., ed. 2, 2: 381 (1963), in Kew Bull. 18: 174 (1964) excl. syn. *Utricularia floridana* Nash, in Fl. Afr. Centr., Lentibulariaceae: 46 (1972); in F.T.E.A., Lentibulariaceae: 19 (1973). TAB. 5. Type from Hispanola.

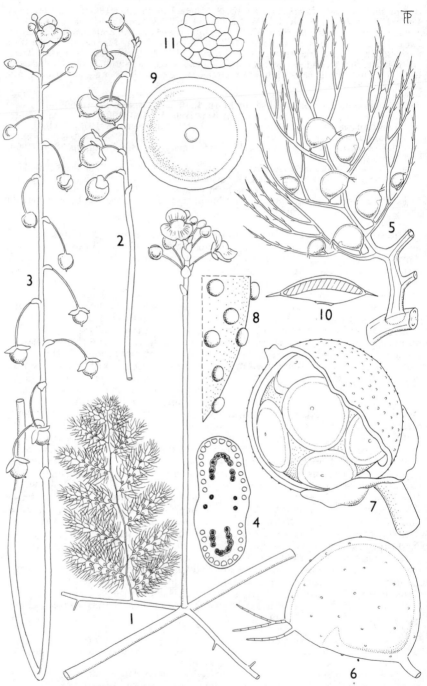

Tab. 5. UTRICULARIA FOLIOSA. 1, part of plant (×1); 2, fruiting peduncle (×1), 1—2 from *Fanshawe* 5752; 3, flowering and fruiting peduncle (×1), *Jordan* 592; 4, cross-section of stolon (×8); 5, part of leaf (×8); 6, trap, lateral view (×30), 4—6 from *Jones* 385; 7, capsule with part of wall removed to show placenta and seeds (×8), *Leonard* 970; 8, capsule wall (×150), *Jones* 385; 9, seed, basal view (×12), *Leonard* 434; 10, cross-section of seed (×12), *Leonard* 970; 11, testa cells (×150), *Jones* 385.

Utricularia foliosa var. *gracilis* Kam. in Engl., Bot. Jahrb. **33**: 111 (1902). Type from Cameroun.

Aquatic herb. Stolons robust, oblong or elliptic in cross-section, smooth and glabrous, 1.3 mm. wide and up to several metres long; internodes 2−15 cm.; rhizoids absent. Leaves broadly ovate in outline, multiple pinnate, up to 15 cm. long, usually dimorphic, some with fewer segments and numerous traps, some with more numerous segments and few or no traps; ultimate segments capillary, minutely setulose. Traps lateral near the base of the penultimate leaf-segments, broadly ovoid, stalked, 1−2 mm. long; mouth lateral, naked or with 2 dorsal simple or sparsely branched hairs. Inflorescence erect, 7−40 cm. high; flowers 3−20, congested at initial anthesis, inflorescence-axis elongating as it matures; peduncle straight, relatively stout, up to 3 mm. thick, smooth and glabrous; scales absent or 1−2 just below the lowermost flower, similar to the bracts; bracts basifixed, broadly ovate or circular, 2−4 mm. long; bracteoles absent; pedicels at first erect, filiform, 4−10 mm. long, elongating and decurving in fruit. Calyx lobes subequal, broadly ovate, 3−4 mm. long, connate at the base, scarcely accrescent; upper lobe with apex subacute, inferior 2−3-dentate. Corolla yellow, 8−15 mm. long; superior lip circular, about twice as long as the upper calyx lobe, with apex rounded; inferior lip larger, oblate to subreniform with apex entire or emarginate; palate much raised, gibbous; spur narrowly conical, straight, more or less two thirds to more or less long as the inferior lip; inner abaxial surface with 2 elliptic patches of shortly stalked glands, one on either side of the central nerve. Filaments linear; anther-thecae confluent. Ovary globose; style short but distinct; stigma inferior lip circular, often ciliate and hispid, superior much smaller, entire or bidentate. Capsule globose, up to 8 mm. in diam., sparsely and minutely glandular, indehiscent. Seeds 4−12, lenticular with a narrow regular wing, 2−2.5 mm. in total diam.; wing 0.2−0.4 mm. wide; testa-cells indistinct, more or less isodiametric.

Botswana. N: Mojeje Dindinga Footpath, fl. 28.iii.1975, Smith 1322 (K; SRGH). **Zambia.** B: Masese, 19.vi.1960, *Fanshawe 5752* (K; SRGH). N: Shiwa Ngandu, 1500 m., fl. 26.vi.1963, *Symoens* 10490 A (K). S: Kafue Flats, Chains Pilot Polder Scheme, 17.vi.1956, *Angus 1336* (EA; K; SRGH).
Very widespread in Africa from Senegal to Angola and S. Africa (Natal), also in Madagascar and in America from N. Carolina to Argentina and the Galapagos. Shallow to deep still or slow-flowing water; from sea level to c. 1200 m.

27. **Utricularia gibba** L., Sp. Pl.: 18 (1753). TAB. 6. Type from U.S.A., (Virginia).
Utricularia obtusa Sw., Prodr. Veg. Ind. Occ.: 14 (1788).—Stapf in Dyer, F.T.A. 4: 495 (1906). TAB. 6. Type from Jamaica.
Utricularia exoleta R. Br., Prodr.: 430 (1810).—Stapf in F.T.A. 4: 495 (1906). Type from Australia.
Utricularia tricrenata Hiern, Cat. Afr. Pl. Welw. 1: 185 (1900). Type from Angola.
Utricularia riccioides A. Chev. in Mém. Soc. Bot. Fr. 8: 187 (1912). Type from Ivory Coast.
Utricularia bifidocalcar Good in Journ. Bot. 62: 161 (1924). Type from Angola.
Utricularia kalmaloensis A. Chev. in Bull. Mus. Hist. Nat. Paris, Sér. 2, 4: 588 (1932). Type from Niger. *Utricularia gibba* subsp. *gibba* P. Taylor in Mitt. Bot. Staatss. München 4: 98 (1961); in F.W.T.A., ed. 2, 2: 38 (1963); in Kew Bull. 18: 198 (1964); in Fl. Afr. Centr., Lentibulariaceae: 49 (1972); in F.T.E.A., Lentibulariaceae: 21 (1973). Type as for *Utricularia gibba*.
Utricularia gibba subsp. *exoleta* (R. Br.) P. Taylor in Mitt. Bot. Staatss. München 4: 101 (1961); in F.W.T.A., ed. 2, 2: 381 (1963); in Kew Bull. 18: 204 (1964); in Fl. Afr. Centr., Lentibulariaceae: 52 (1972); in F.T.E.A., Lentibulariaceae: 22 (1973). Type as for *Utricularia exoleta*.

Aquatic or subaquatic herb. Stolons usually up to more or less 10, fasciculate at the base of the peduncle, filiform, terete, up to 20 cm. × 1 mm., sometimes slightly inflated. Rhizoids few from the base of the peduncle, often absent, filiform. Leaves numerous, alternate on the stolons, up to 20 mm. long; segments capillary, glabrous or sparsely setulose, simple or usually dichotomously forked from or near the base, each segment sometimes again 1−3 or rarely more times forked. Traps usually numerous, replacing one of the leaf segments at a fork, ovoid, 1−1.5 mm. long, stalked; mouth lateral, with 2 dorsal, long, usually copiously branched hairs and a few ventral shorter simple hairs. Inflorescence erect, 2—35 cm. high, solitary or often several arising in succession from the fascicle of stolons; flowers (1)2(6); peduncle straight, smooth and glabrous; scales absent or more usually 1 at or above the middle of the peduncle, similar to the bracts; bracts basifixed, semi-circular, semi-amplexicaul, more or less 1 mm. long; bracteoles absent; pedicels filiform, erect, (2)6—12(30) mm. long. Calyx lobes subequal, circular to broadly ovate, 1—3 mm. long. Corolla yellow, often with brown or reddish nerves, 4—25 mm. long; superior lip circular, 2—3 times as long as the upper calyx lobe, usually shallowly and obscurely 3-lobed; inferior lip usually shorter and narrower than the superior, circular, usually entire, rarely emarginate; palate much raised and bigibbous; spur conical to narrowly cylindrical, shorter than to almost twice as long as the inferior lip, the apex usually bearing a few shortly stalked glands. Filaments linear; anther-thecae more or less confluent. Ovary globose; style short but distinct; stigma inferior lip semi-circular, superior short, more or less obsolete. Capsule globose, 2—4 mm. in diam., dehiscing into 2 lateral valves. Seeds rather few (more or less 20—40), imbricate on the smooth placenta, lenticular with a broad irregular wing, 1—1.6 mm. in total diam., distal surface smooth to verrucose; hilum prominent; testa-cells small, irregular, indistinct.

Botswana. N: Botletle R., 1.6 km. S. of Samadupi Drift, 1000 m., fl. 23.i.1972, *Gibbs-Russell* 1397 (K; SRGH). **Zambia**. B: 4.8 km. S. of Kalabo, 16.xi.1959, *Drummond & Cookson* 6538 (K; MO; SRGH). N: Mbala, Lake Chila, 1500 m., 29.viii.1956, *Richards* 6007 (K). W: Mwinilunga Distr., Kalenda Plain, 18.xi.1937, *Milne-Redhead* 3310 (K). C: Lusiwasi, 20.ii.1960, *Richardson & Livingstone* s.n. (NC USA; K). E: Chadiza, 850 m., 1.xii.1958, *Robson* 80 (BR; EA; K; LISC; SRGH). S: Chepezami Dam, 22 km. W. of Pemba, 1380 m., fl. 21.iv.1954, *Robinson* 707 (K). **Zimbabwe**. N: Urungwe, Zwipani, 1050 m., 12.x.1957, *Phipps* 782 (K; SRGH). W: Hwange Distr., Matetsi Safari Area Unit, 13.ix.1979, *Gonde* 251 (K; SRGH). C: Harare, Makabusi R., 1470 m., 21.ix.1947, *Wild* 2012 (K; SRGH). Malawi. N: Mzimba Distr., 16 km., SW. of Mzuzu, Mbowe Dam, 1350 m., 2.viii.1970, *Pawek* 3668 (K). **Mozambique**. T: Baroma Distr., fl. 11.vii.1950, *Chase* 2601 (BM). MS: Biera, Gorongosa National Park, Xirulo 60—70 m., fl. 6.vii.1972, *Ward* 7759 (K; Durban Coll). M: Maputo, Matutuino fl. 3.viii.1980, *de Koning* 7854 (BM; LIJSC).

Widespread in Africa from Senegal & Ethiopia to S. Africa and Madagascar, also throughout the tropics and subtropics, in N. America and as a weed of botanic gardens throughout the world. Shallow water and mud by ditches streams and lakes; sea level to 2460 m.

28. **Utricularia cymbantha** Oliver in Journ. Linn. Soc., Bot. 9: 147 (1865).—Stapf in Dyer, F.T.A. 4: 494 (1906).—P. Taylor in Kew Bull. 18: 209 (1964); in Fl. Afr. Centr., Lentibulariaceae: 52 (1972); in F.T.E.A., Lentibulariaceae: 22 (1973). Type from Angola.
    Biovularia cymbantha (Oliver) Kam. in Engl., Bot. Jahrb. 33: 113 (1902). Type as above.
    Utricularia stephensiae Lloyd, Carn. Pl. 213, 214, 224, t.23/12 & 13 (1942), nomen nudum. Type: Mozambique (not seen).
    Sacculina madecassa Bosser in Nat. Malgache 8: 27 (1956). Type from Madagascar.

Aquatic herb. Stolons capillary, up to 10 cm. long or more × c. 0.2 mm., minutely glandular; rhizoids absent. Leaves numerous, forked from the base into 2 equal or unequal capillary segments, 1—2.5 mm. long; internodes 1—2 mm. long. Traps numerous, inserted in the angle between the leaf-segments or laterally on the longer leaf-segment, ovoid, stalked, 1—1.5 mm. long; mouth lateral, with 2 dorsal capillary more or less branched hairs and 3 shorter ventral simple hairs. Inflorescence lateral, erect; peduncle capillary, ovoid, more or less 1 mm. long; bracteoles absent; pedicel capillary, cernuous in bud and in fruit, erect at anthesis, 1.5—4 mm. long. Calyx lobes subequal, circular, more or less 1 mm. long, accrescent. Corolla white or cream, 3—5 mm. long; superior lip 1.5—2 times as long as the upper calyx lobe, semi-circular with apex entire or emarginate; inferior lip circular, the base subcordate and with apex rounded, entire or emarginate; palate scarcely raised, minutely glandular; spur very short, saccate. Filaments capillary; anther-thecae confluent. Ovary ellipsoid, more or less 0.4 mm. long; ovules 2—5; style distinct, filiform, as long as the ovary; stigma inferior lip circular, ciliate, superior obsolete. Capsule ellipsoid, more or less 1 mm. in diam., the wall membranous, apparently indehiscent. Seeds

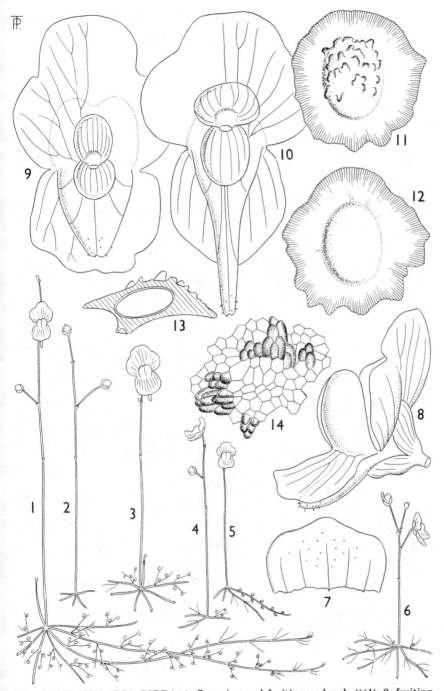

Tab. 6. UTRICULARIA GIBBA. 1, flowering and fruiting peduncle (×1); 2, fruiting peduncle (×1), 1—2 from *Milne-Redhead & Taylor* 8076; 3, flowering peduncle (×1), *Williams* in E.A.H. 12347; 4 & 5, flowering and fruiting or flowering peduncles (×1), 4—5 from *Richards* 12875; 6, flowering and fruiting or flowering peduncles (×1), *Richards* 6089; 7, bract, flattened (×30), *Drummond & Hemsley* 4642; 8, flower, lateral view (×8); 9, flower, adaxial view (×8), *Milne-Redhead & Taylor* 8076; 11 & 12, two seeds from the same capsule (×30); 13, cross-section of seed (×30); 14, testa cells from verrucose seed (×150), 11—14 *Richards* 12875.

2 or 3, lenticular, narrowly and regularly winged, total diam., more or less 0.7 mm.; testa cells indistinct, irregular, more or less elongated.

Botswana. N: 240 km. N. of Maun, 7.vii.1937, *Erens* 351 (K; PRE). **Zambia.** N: Mbala, Lake Chila, 1650 m., 15.iv.1963, *Richards* 18087 (K). **Mozambique.** Specimen unlocalised and not seen.

Also from Zaire, Uganda, Angola, S. Africa (Transvaal) and Madagascar. Floating in shallow water; 900−1800 m.

2. GENLISEA A. St. Hil.

Genlisea A.St. Hil., Voy. Dist. Diam. 2: 428 (1833).—Stapf in Dyer, F.T.A. 4: 497 (1906).—P. Taylor in F.W.T.A., ed. 2, 2: 375 (1963); in Fl. Afr. Centr., Lentibulariaceae: 53 (1972); in F.T.E.A., Lentibulariaceae: 23 (1973).

Rootless perennial or annual herbs of wet places. Stem short, subterranean, erect or decumbent. Leaves dimorphic, persistent at anthesis; foliage leaves petioled, entire, linear-lanceolate to spathulate or circular, glabrous or rarely hairy, densely or laxly rosulate from the upper part of the stem; pitcher leaves (traps) more or less densely congested on the lower part of the stem and decending into the substrate and consisting of a stalk and a slender tube, cylindrical from an ellipsoid base and terminating in two ribbon-like helically twisted arms and tube provided internally with transverse rows of stiff inwardly directed hairs. Inflorescence terminal and arising from the leaf-rosette, racemose, bracteate; peduncle simple or branched above, erect, usually glandular or hispid, rarely completely glabrous, provided with more or less numerous sterile bracts (scales); raceme congested to more or less elongated, few-many-flowered; pedicels usually considerably longer than the bracts, erect at anthesis, erect, spreading or strongly decurved in fruit, glandular, hispid or glabrous. Bracts basifixed; bracteoles 2, inserted with the bract at the base of the pedicel. Calyx lobes 5, subequal, slightly accrescent, densely glandular to hispid or more or less glabrous. Bracts basifixed; bracteoles 2, inserted with the bract at the base of the pedicel. Calyx lobes 5, subequal, slightly accrescent, densely glandular to hispid or glabrous. Corolla bilabiate, glandular, hispid or glabrous, blue, violet, mauve, yellow or white; superior lip more or less erect, entire or bilobed; inferior lip larger, spurred at the base; palate raised and more or less gibbous, limb spreading or deflexed, more or less deeply 3-lobed; spur acute or obtuse, shorter than to longer than the inferior lip. Stamens 2, inserted at the base of the corolla; filaments falcate; anthers dorsifixed, ellipsoid, the thecae more or less confluent. Ovary globose, glandular, hispid or glabrous, style short, indistinct; stigma bilabiate; inferior lip about as broad as the ovary, semi-circular; superior lip more or less obsolete; ovules numerous, sessile on a fleshy free basal placenta, anatropous. Capsule globose, valvate or uniquely multiple-circumscissile, the lower most line of dehiscence approximately equatorial and with two others between it and the persistent style. Seeds exalbuminous, numerous, ovoid, reticulate, with a prominent hilum at one end.

A small genus of 18 species occuring in S. and C. America (10 species), tropical Africa and Madagascar.

1. Fruiting pedicel strongly decurved; scales on peduncle numerous -   -   - 2
   − Fruiting pedicel erect; scales on peduncle few -   -   -   -   -   -   - 4
2. Corolla cream; flowers fewer than 6, not congested -   -   -   - . 1. *pallida*
   − Corolla mauve or violet; flowers usually 10 or more, congested   -   -   - 3
3. Inflorescence 6−14 cm. high, densely glandular throughout . 2. *glandulosissima*
   − Inflorescence 20−50 cm. high, densely glandular above, glabrous or only sparsely glandular below -   -   -   -   -   -   -   -   -   - 3. *margaretae*
4. Inflorescence glandular   -   -   -   -   -   -   -   -   - 4. *africana*
   − Inflorecsence not glandular   -   -   -   -   -   -   -   - 5
5. Ovary and capsule densely hispid, at least in the upper part -   - 5. *hispidula*
   − Ovary and capsule glabrous or with a very few hairs in the upper part
     -   -   -   -   -   -   -   -   -   -   -   -   - 6. *subglabra*

1. **Genlisea pallida** P. Taylor & E. Fromm-Trinta in Bradea 4 (27): 176 (1985). Type: Zambia, Mwinilunga Distr., Sinkabolo Dambo, *Milne-Redhead* 3575 (K, holotype; R, isotype).

Terrestrial herb. Stem short, erect. Leaves very numerous, densely rosulate, spathulate, up to 3 cm. × 4 mm. Traps numerous, up to 10 cm. total length. Inflorescence erect, simple, up to 15 cm. high; peduncle straight, rigid, terete, c. 0.7 mm. thick, sparsely covered with short gland-tipped hairs below, more densely so above; flowers 4−8, distant; scales numerous, similar to the bracts; bracts basifixed, ovate-lanceolate, c. 1.5 mm. long; bracteoles similar but shorter; pedicels erect at anthesis, elongating and strongly decurving in fruit, 5−10 mm. long, more or less densely covered with short gland-tipped hairs. Calyx lobes oblong-ovate, obtuse, subequal, c. 1.5 mm. long, 3-nerved, more or less densely covered externally with short gland-tipped hairs. Corolla cream with a yellow palate, 7−10 mm. long with a few gland-tipped hairs on the spur, glabrous or with sparse sessile glands elsewhere; superior lip ovate-oblong, with apex truncate, about twice as long as the calyx lobes; inferior lip longer and broader, rather shallowly 3-lobed; palate slightly raised; spur cylindrical from a conical base, 2−3 times as long as the inferior lip. Filaments falcate, 1 mm. long; anther-thecae subdistinct. Ovary globose, densely covered with short gland-tipped hairs; style short; stigma inferior lip semi-circular, the upper lip obsolete. Capsule globose, c. 3 mm. in diam., densely covered with short gland-tipped hairs. Seeds numerous, ovoid, c. 0.3 mm. long, reticulate; testa cells more or less isodiametric.

Zambia. W: Sinkabolo Dambo, fl. 9.xii.1937, *Milne-Redhead* 3575 (K, holotype; R, isotype).
Also in Angola. In permanently wet peat bogs, flowering in the wet season.

2. **Genlisea glandulosissima** R.E. Fries in Wiss. Ergebn. Schwed. Rhod.-Kongo-Exped. 1: 301 (1914). Type: Zambia, Nsombo, Lake Bangweulu, *Fries* 1050 (UPS, holotype).

Terrestrial herb. Stem short, erect. Leaves very numerous, densely rosulate, spathulate, up to 2.5 cm. × 5 mm. Traps numerous, up to 5 cm. long. Inflorescence erect, simple, 6−14 cm. high; peduncle straight, rigid, terete, 1−1.5 mm. thick, densely covered with a mixture of short and long gland-tipped hairs throughout; flowers 10−20, congested; scales numerous, similar to the bracts but c. twice as long; bracts basifixed, ovate-lanceolate, 1.5 mm. long; bracteoles similar but shorter; pedicels spreading and 3−5 mm. long at anthesis, elongating and strongly decurving in fruit, more or less densely covered with long and short gland-tipped hairs. Calyx lobes oblong-ovate, obtuse, subequal, c. 2 mm. long, 3-nerved, more or less densely covered externally with gland-tipped hairs. Corolla mauve or purple, 6−8 mm. long, covered externally with short gland-tipped hairs; superior lip ovate-oblong, with apex truncate, less than twice as long as the calyx lobes; inferior lip longer and broader, distinctly 3-lobed; palate slightly raised; spur cylindrical from a conical base, about twice as long as the inferior lip. Filaments falcate, c. 1 mm. long; anther-thecae subdistinct. Ovary globose, densely covered with long gland-tipped hairs; style short; stigma inferior lip semi-circular, the superior lip obsolete. Capsule globose, 2.5−3 mm. in diam., densely covered with long gland-tipped hairs. Seeds numerous,, ovoid, 0.3−0.4 mm. long, reticulate; testa cells more or less isodiametric.

Zambia. N: Shiwa Ngandu, 1620 m., fl. 19.vii.1938, *Greenway* 5423 (K).
Permanently wet peaty bogs, flowering in the dry season; 1200−1500 m.

3. **Genlisea margaretae** Hutch., Botanist in S. Afr.: 529 (1906).−P. Taylor in F.T.E.A., Lentibulariaceae: 23 (1973). TAB. 7 fig. C. Type: Zambia, N. of Kasama, *Hutchinson & Gillet* 4050 (K, holotype; BM, isotype).
   Genlisea recurva Bosser in Le Naturaliste Malgache 10: 23 (1958). Type from Madagascar.

Terrestrial herb. Stem short, erect. Leaves very numerous, densely rosulate, spathulate, up to 3 cm. × 4 mm. Traps numerous, up to 20 cm. total length. Inflorescence erect, simple, 20−35(50) cm. high; peduncle straight, rigid, terete, 1−1.5 mm. thick, sparsely or very sparsely covered with short gland-tipped hairs below, more densely so above; flowers (4)6−10(20), congested; scales numerous, similar to the bracts but somewhat longer; bracts basifixed, ovate-lanceolate, c. 2 mm. long; bracteoles similar but shorter; pedicels suberect at anthesis, 1—3 mm. long, elongating and strongly decurving in fruit, more or less densely covered with short gland-tipped hairs. Calyx-lobes oblong-ovate, obtuse, subequal, c. 2 mm. long, 3-nerved, more or less densely covered externally with short gland-tipped hairs. Corolla mauve or purple, 7—10 mm. long, covered externally with short gland-tipped hairs; superior lip ovate-oblong, with apex truncate, about twice as long as the calyx lobes; inferior lip longer and broader, rather shallowly 3-lobed; palate slightly raised; spur cylindrical from a conical base, 2—three times as long as the inferior lip. Filaments falcate c. 1 mm. long; anther-thecae subdistinct. Ovary globose, densely covered with short gland-tipped hairs; style short; stigma inferior lip semi-circular, the superior lip obsolete. Capsule globose, 2.5—3 mm. in diam., densely covered with short gland-tipped hairs. Seeds numerous, ovoid, 0.3—0.4 mm. long, reticulate; testa cells more or less isodiametric.

Zambia. N: Kasama, fl. 1.vii.1964, *Fanshawe* 8778 (K; NDO).
Also in Tanzania and Madagascar. Permanently wet peaty or sandy bogs, flowering in the dry season; 1200—1500 m.

4. **Genlisea africana** Oliver in Journ. Linn. Soc., Bot. 9: 145 (1865).—Stapf in Dyer, F.T.A. 4: 497 (1906).—P. Taylor in F.W.T.A., ed. 2, 2: 375 (1963) excl. syn. *Genlisea stapfii*. TAB. 7 fig. B. Type from Angola.
Genlisea africana forma *pallida* R.E. Fries in Wiss. Ergebn. Schwed. Rhod.-Kongo-Exped. 1: 301 (1916). Type from Zaire, Katanga.
Genlisea subviridis Hutch., Botanist in S. Afr.: 528 (1944). Type: Zambia, 96 km. NE. of Kabwe, *Hutchinson & Gillet* 3644 (K, holotype).
Genlisea africana subsp. *africana* P. Taylor in Fl. Afr. Centr., Lentibulariaceae: 57 (1972). Type as for *Genlisea africana*.

Terrestrial herb. Stem short, erect. Leaves usually numerous, rosulate, spathulate, usually long-petiolate, (1)2—4(5) cm. × up to 9 mm. Traps about as numerous as the leaves. Inflorescence erect, simple or sparsely branched above, (5)10—30(50) cm. high; peduncle terete, sparsely to densely covered with long gland-tipped hairs above, more or less glabrous below; flowers (1)3—10 or more on branched inflorescences; scales few, similar to the bracts; bracts ovate-lanceolate to linear-lanceolate, acute or acuminate, 2—4 mm. long with marginal gland-tipped hairs; bracteoles similar but narrower; pedicels erect, (2)5—25(30) mm. long at anthesis usually elongating in fruit, more or less densely covered with long gland-tipped hairs. Calyx lobes subequal, ovate-lanceolate, acute or acuminate, 2—4 mm. long, densely covered with long gland-tipped hairs. Corolla violet, blue, mauve or pink with a greenish or yellowish spur or sometimes entirely yellow, (5)10—12(18) mm. long, externally sparsely covered with short gland-tipped hairs; superior lip broadly ovate, about three as long as the calyx lobes; inferior lip larger, broader than long, 3-lobed; palate raised, gibbous; spur cylindrical, obtuse, c. 1.5 times as long as the inferior lip. Filaments falcate; anther-thecae subdistinct. Ovary more or less densley covered with short gland-tipped hairs; style short, indistinct; stigma inferior lip semi-circular, superior obsolete. Capsule globose to braodly ovoid, 3—4 mm. long, densely covered with gland-tipped hairs. Seeds numerous, ovoid, 0.4—0.5 mm. long, conspicuously reticulate; testa cells more or less isodiametric.

Zambia. N: Chishinga Ranch near Luwingu, 1380 m., fl. 27.iv.1961, *Astle* 546 (K). W: Chingola, Luano Forest Reserve., c. 1350 m., fl. 27.viii.1961, *Linley* 176 (K; SRGH). C: Mkushi R. Hotel, fl. 23.vi.1963, *Robinson* 5536 (K). S: 8 km. E. of Choma, 1290 m., fl. 27.iii.1955, *Robinson* 1187 (K; SRGH).**Zimbabwe**. W: Matobo, Farm Besna Kobila, 1440 m., fl. vii.1955, *Miller* 2940 (K; SRGH). C: Harare, fl. 20.ix.1953, *Wild* 4140 (K; SRGH).
Also in Zaire and Angola. Marshes and wet grassland; 1200—1500 m.

5. **Genlisea hispidula** Stapf in Dyer, F.C. 4: 437 (1904); F.T.A., 4: 498 (1906).—P. Taylor in F.W.T.A., ed. 2, 2: 375 (1963). Type from S. Africa.
Genlisea hispidula subsp. *hispidula* P. Taylor in F.T.E.A., Lentibulariaceae: 25 (1973). Type as for *Genlisea hispidula*.

Terrestrial herb. Stem short, erect. Leaves usually few, laxly to quite densely rosulate, spathulate, usually long-petiolate, (1)2—4(5) cm. × up to 9 mm. wide. Traps about as numerous as the leaves. Inflorescence erect, simple or sparsely

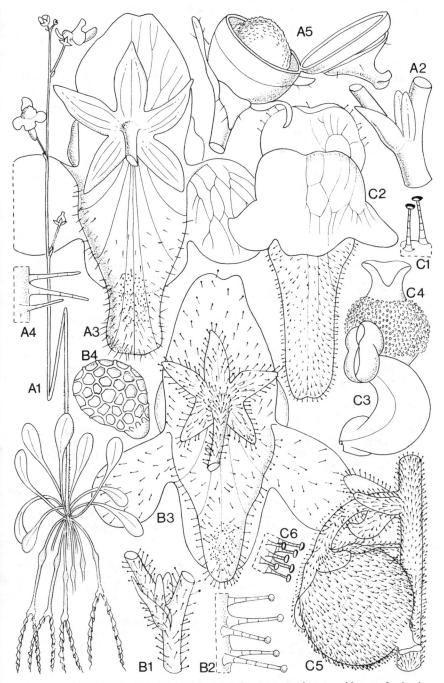

Tab. 7. A.—GENLISEA SUBGLABRA. A1, habit (×1); A2, bract and bracteoles in situ (×8); A3, flower, back view (×8); A4, hairs on spur (×60), A1–4 from *Milne-Redhead & Taylor* 8009; A5, dehisced capsule (×8), from *Richards* 967. B.—GENLISEA AFRICANA. B1, bract and bracteoles in situ (×8); B2, glands on pedicel (×60); B3, flower, back view (×8); B4, seed (×60), all from *Richards* 9995. C.—GENLISEA MARGARETAE. C1, glands on pedicel (×48); C2, flower, front view (×8); C3, stamen (×24); C4, pistil (×24); C5, capsule in situ (×8); C6, glands on capsule wall (×48), all from *Milne-Redhead & Taylor* 10013a.

branched above, (6)10—25(30) cm. high; peduncle flexuous, terete, glabrous above, usually sparsely hispidulous at or near the base only; flowers (1)3—10(15); scales few, similar to the bracts; bracts ovate-lanceolate to linear-lanceolate, acute or acuminate, 2—4 mm. long, (2)5—15(25) mm. long at anthesis, usually elongating in fruit, glabrous throughout or distally hispidulous. Calyx lobes subequal, ovate-lanceolate to lanceolate, acute or acuminate, 2—4 mm. long, sparsely to densely hispidulous. Corolla violet, blue, mauve or pink with a greenish or yellowish spur or rarely wholly yellow, (6)10—12(15) mm. long; superior lip ovate, about twice as long as the calyx lobes; inferior lip larger, broader than long, 3-lobed; palate raised, gibbous; spur cylindrical, obtuse, 1.5—2 times as long as the inferior lip, more or less hispidulous. Filaments falcate; anther-thecae subdistinct. Ovary ovoid, more or less densely hispidulous; style short, indistinct; stigma inferior lip semi-circular, superior obsolete. Capsule globose to broadly ovoid, 3—4 mm. long, more or less densely hispidulous. Seeds numerous, ovoid, 0.4—0.5 mm. long, conspicuously reticulate; testa cells more or less isodiametric.

Zambia. W: c. 160 km. NE. of Dobeka Bridge, *Milne-Redhead* 3607 (K). Zimbabwe. E: c. 3 km. a pedibus montis, c. 2000 m., fl. 3.xii.1930, *Fries, Norlindh & Weimark* 3444 (K). C: Marondera, fl. 20.iv.1948, *Corby* 99 (K; SRGH). Malawi. N: Nyika Plateau, 12.5 km. from Chelinda Camp on road to Ngandu, just after the turn off to Rukuru Bridge, 2200 m., fl. 11.iii.1977, *Grosvenor & Renz* 1134 (K; SRGH). S: Zomba Mt., 1850 m., fl. 25.i.1959, *Robson & Jackson* 1320 (K). Mozambique. MS: Chimanimani Mts., c. 3.2 km., SE. of the saddle between St. George's and Poacher's Cave, 1500 m., fl. 12.iv.1967, *Grosvenor* 400 (K; SRGH).
Also in Nigeria, Cameroun, Central African Republic, Kenya, Tanzania and S. Africa. Seasonally or permanently wet grassland; 1000—2700 m.

6. **Genlisea subglabra** Stapf in Dyer, F.T.A., 4: 498 (1906). TAB. 7 fig. A. Type: Zambia, Fwambo, *Nutt* s.n. (K, holotype).
Genlisea hispidula subsp. *subglabra* (Stapf) P. Taylor in Kew Bull. 26: 444 (1972); in Fl. Afr. Centr., Lentibulariaceae: 58 (1972); in F.T.É.A., Lentibulariaceae: 25 (1973).

Terrestrial herb. Stem short, erect. Leaves usually few, laxly rosulate, spathulate, usually long-petiolate, 1—5 cm. × up to 9 mm. Traps about as numerous as the leaves. Inflorescence erect, simple or sparsely branched above, 10—25(40) cm. high; peduncle flexuous, terete, glabrous above, usually sparsely hispidulous at or near the base only; flowers 3—10; scales few, similar to the bracts; bracts ovate-lanceolate to linear-lanceolate, acute or acuminate, 2—4 mm. long, glabrous or sparsely hispidulous above. Calyx lobes subequal, ovate-lanceolate to lanceolate, acute or acuminate, 2—4 mm. long, sparsely hispidulous or glabrous. Corolla violet, blue, mauve or pink, rarely yellow or white, externally sparsely hispidulous, 10—15 mm. long; superior lip ovate, about twice as long as the calyx lobes; inferior lip larger, broader than long, 3-lobed; palate raised, gibbous; spur cylindrical, obtuse, 1.5—2 times as long as the inferior lip. Filaments falcate; anther-thecae subdistinct. Ovary ovoid, glabrous or very sparsely hispidulous; style short, indistinct; stigma inferior lip semi-circular, superior obsolete. Capsule globose to broadly ovoid, 3—4 mm. long, glabrous or very sparsely hispidulous. Seeds numerous, ovoid, 0.4—0.5 mm. long, conspicuously reticulate; testa cells more or less isodiametric.

Zambia. N: Mbala, Old Mbala-Sumbawanga Rd., 16 km. from Kawimbe, 1620 m., fl. 7.ix.1956, *Richards* 6128 (K). Malawi. N: Nkhata Bay Distr., Vipya Plateau, 1650 m.,fl. 11.x.1973, *Pawek* 7492 (K; MO; SRGH; UC). S: Zomba Distr., Zomba Plateau, 1830 m., fl. 4.v.1970, *Brummitt & Banda* 10347 (K).
Also in Burundi, Zaire and Tanzania. Seasonally or permanently wet grassland; 1000—1500 m.

123. GESNERIACEAE

By O.M. Hilliard and B.L. Burtt

Acaulescent or caulescent herbs, or rarely shrubs. Leaves opposite (rarely alternate), those of a pair equal or unequal; plants sometimes unifoliate and the leaf cotyledonary in origin. Inflorescence generally of open axillary cymes, the flowers at each dichotomy being paired and opening serially; occasionally much congested and sub-capitate, or pseudoracemose. Flowers hermaphrodite (very rarely unisexual) often large and showy, sometimes cleistogamous with reduced corolla. Calyx tubular and 5-lobed or divided to the base or 3 upper lobes only united. Corolla gamopetalous with distinct tube, often bilabiate, proportion of lobes to tube variable; lobes imbricate, and adaxial pair often inside. Stamens usually 4 or 2, rarely 5, inserted on corolla tube; anthers free or variously connate, 2-celled, opening longitudinally. Disk annular or cup-like, often lobed or undulate, rarely oblique or absent. Ovary superior, unicellular with 2 parietal bilamellate placentas, occasionally bilocular by their union centrally. Ovules numerous. Fruit a capsule (often linear, sometimes spirally twisted) or a more or less fleshy berry. Seeds numerous, small, more or less ellipsoid, sometimes tailed with hair-like appendages at either end; endosperm absent or very slight.

Mainly in the tropics and subtropics, often favouring shady places (at least for their roots) and sometimes epiphytic. Eight genera in Africa. The African genera all belong to subfam. *Cyrtandroideae* and either occur in Asia or have affinity with genera there. They are not closely related to the tropical American genera.

The family description does not include features peculiar to the New World genera of subfamily *Gesnerioideae*.

STREPTOCARPUS Lindl.

Streptocarpus Lindl. in Bot. Reg. 14 tab. 1173 (1828).—Hilliard & Burtt, Streptocarpus: An African Plant Study (1971).

Herbs, rarely subshrubs; caulescent, erect or creeping, or acaulescent, rosulate or unifoliate; annual, perennial or monocarpic. Leaves opposite and petiolate in the caulescent species (very rarely alternate and sessile, and then the flowering axis with a very large basal leaf); in the acaulescent species the leaf may show continued growth from a basal meristem, evidenced by wide-spreading lateral veins and frequent absence of intact apex; indumentum of simple glandular or eglandular hairs, rarely of branched hairs; sessile or subsessile glands usually present: the lower surface of the leaf sometimes white-dotted due to the presence of stomatal turrets. Inflorescences axillary or apparently from base of lamina or from leaf-stalk; cymose, branching, sometimes more or less one-sided, the flowers paired, rarely reduced to two flowers or only one. Calyx usually divided to the base into 5 segments, more rarely with a distinct tube and 5 teeth. Corolla gamopetalous, 5-lobed, most often distinctly bilabiate. Stamens arising at various levels on the corolla tube, the anterior 2 only fertile; lateral staminodes usually present, the posterior one often missing; filaments variable in length, often thickened in the middle; anthers with divaricate, rarely parallel, lobes, the lines of dehiscence confluent at the apex, usually cohering face to face. Disk annular or shortly cupular. Ovary usually unilocular, but sometimes apparently bilocular by fusion of the T-shaped intrusive placentas; ovules restricted to the recurved tips of the placentae; ovary narrowed, often very gradually, into the style; stigma variable. Capsule more or less cylindrical, varying from short and broad (5 mm.) to long and slender (180 mm.), twisting spirally before maturity, dehiscing by a loculicidal slit and slight untwisting of the spiral, when old sometimes splitting into 4 valves. Seeds numerous, small, with very little endosperm, reticulate or verruculose. Seedlings with the cotyledons becoming unequal after germination.

A genus of about 135 species in Africa, Madagascar and E. Asia, but the four Asiatic species are doubtfully congeneric.

Streptocarpus is divisible into two subgenera: subgen. *Streptocarpus* comprising acaulescent species (there are a few exceptions), which develop an abscission layer towards the base of the lamina during the unfavourable season and have the peduncles arising from the base of the lamina or from the petiolode; and subgen. *Streptocarpella* K. Fritsch, comprising caulescent species, which lack radical leaves; the cauline leaves have no abscission layer and the peduncles are axillary. The subgenera are also separated by a difference in chromosome number: 2n = 32 in subgen. *Streptocarpus*, 2n = 30 in subgen. *Streptocarpella*. subgen. *Streptocarpella* is represented in the area of Flora Zambesiaca only by *S. buchananii*; all the other species in this area belong to subgen *Streptocarpus* and are strictly acaulescent with the exception of *S. myoporoides*.

In a revision of the whole genus (Hilliard & Burtt 1971) an aggregate species concept was employed where necessary to give a loose linkage between very closely allied species whose reduction to subspecific rank would scarcely have been warranted and would have led to an unnecessarily cumbersome nomenclature.

Two such aggregate species are represented in the Flora Zambesiaca area. *S.* agg. *cooperi* (based on the Natal species *S. cooperi* C.B.Cl.) includes from our area *S. grandis*, *S. michelmorei* and *S. solenanthus*; *S.* agg. *monophyllus* (based on the Angolan *S. monophyllus* Welw.) includes *S. eylesii*, *S. wittei* and *S. arcuatus*.

This aggregate species concept is not included in the formal presentation of the Flora account, but it is as well to remind the user that it has been found necessary. In some cases the component species of these aggregates may not be easy to distinguish.

The occurrence of wild hybrids may be a source of difficulty in naming specimens, particularly in the eastern highlands of Zimbabwe and in adjoining Mozambique (see Hilliard & Burtt 1971, pp. 91-92). Hybrids are likely both within and between the two aggregate species mentioned above; otherwise *S. umtaliensis* is the only species that is at all likely to be involved and the evidence here is very slight. Individual instances are mentioned under the species concerned. It can only be emphasised that when a plant or population seems to vary from the typical form of its species an attempt to relate this to characters of other species present in the area is always worth while.

1. Stemmed herbs with opposite leaves · · · · · · · · · 2
 — Acaulescent herbs, leaves solitary or in a rosette or tuft · · · · · 3
2. Inflorescence axillary, diffuse; corolla blue, mouth almost closed, limb with upper lip bilobed, lower 3- lobed · · · · · · · · 22. *buchananii*
 — Inflorescence terminal, condensed; corolla red, mouth open, limb with upper lip 4-lobed, lower 1-lobed · · · · · · · · · 9. *myoporoides*
3. Corolla mouth strongly compressed from the sides · · · · · · 4
 — Corolla mouth open · · · · · · · · · · · 6
4. Plant with several densely sericeous leaves; lateral veins forked at the apices and closely outlining the deeply cut leaf margin · · · · 13. *nimbicola*
 — Plant with only one thinly pilose leaf; leaf margin crenulate, not outlined by strong veins · · · · · · · · · · · · · · 5
5. Stigma capitate with a median transverse groove; corolla up to 35 mm. long, tube up to 22 mm. long, inferior lip 15−20 mm. broad; hairs on ovary spreading · · · · · · · · · · · · · 4. *goetzei*
 — Stigma distinctly bilobed; corolla 35−55 mm., tube usually more than 25 mm. long, lower lip up to 30 mm. broad; hairs on ovary appressed
· · · · · · · · · · 5. *confusus* subsp. *lebomboensis*
6. Corolla hooded, lateral lobes forming part of the upper lip · 9. *myoporoides*
 — Corolla not hooded, lateral lobes forming part of the lower lip · · · · 7.
7. Corolla c. 85 mm. long · · · · · · · · · 11. *dolichanthus*
 — Corolla up to 40 mm. long · · · · · · · · · · · 8
8. Floor of corolla tube and base of lower lip patterned with violet or magenta-pink spots and/or stripes · · · · · · · · · · · 9
 — Floor of corolla tube and base of the lower lip never patterned with violet or magenta-pink spots and/or stripes · · · · · · · · · · 17
9. Plants with several leaves of similar size arranged in a rosette · · 10
 — Plants either with 1 leaf only, or with 1 or 2 large leaves loosely associated with 1 or more smaller ones · · · · · · · · · · · 13
10. Leaves flannelly with whitish hairs; corolla tube dull reddish-violet, spotted inside, limb white · · · · · · · · · · 21. *rhodesianus*
 — Leaves hairy but not white-flannelly · · · · · · · · 11
11. Ovary glabrous or with a few sessile glands · · · · 20. *leptopus*
 — Ovary pubescent · · · · · · · · · · · · 12
12. Hairs on ovary eglandular; corolla not swollen above base · 10. *milanjianus*
 — Hairs on ovary glandular; corolla swollen above base on upper side
· · · · · · · · · · · · · 12. *hirtinervis*

13. Plant with one sessile leaf - - - - - - - 16. *erubescens*
  − Plant usually with two or more leaves, the flowering leaf with a conspicuous stalk (petiolode) - - - - - - - - - - - - - 14
14. Corolla tube broadly subcylindrical, c. 7 mm. long, 5 mm. - 14. *brachynema*
  − Corolla tube funnel-shaped, c. 9−25 mm. long - - - - - - - 15
15. Tube c. 9 mm. long, 2−3 mm. across mouth - - - - - - 16
  − Tube c. 18−25 mm. long 6−7 mm. across mouth - - - 17. *cyanandrus*
16. Capsule almost glabrous - - - - - - - - 18. *pumilus*
  − Capsule hirsute - - - - - - - - - - 19. *hirticapsa*
17. Ovary hairs appressed; corolla tube up to 10 mm. long - - - - 18
  − Ovary hairs spreading or half-spreading; corolla tube usually exceeding 10 mm. - - - - - - - - - - - - - - 19
18. Corolla c. 11−13 mm. long, usually pure white, very occasionally blue - - - - - - - - - - - - 15. *umtaliensis*
  − Corolla 15−17 mm. long, pale violet with a median yellow bar on floor of tube - - - - - - - - - - - 14. *brachynema*
19. Lower lip up to 15 mm. long; flowers scentless - - - - - - 20
  − Lower lip more than 15 mm. long; flowers smelling strongly of creosote and honey - - - - - - - - - - - - 22
20. Ovary hairs half-spreading - - - - 1. *grandis* subsp. *septentrionalis*
  − Ovary hairs spreading - - - - - - - - - - - 21
21. Corolla richly coloured, tube exceeding 25 mm., lower lip 10 mm. or longer - - - - - - - - - - - 2. *michelmorei*
  − Corolla usually very pale, tube up to 25 mm. long, lower lip 5−8 mm. long - - - - - - - - - - - 3. *solenanthus*
22. Corolla tube swollen on the upper side at the base and strongly arcuate - - - - - - - - - 8. *arcuatus*
  − Corolla tube not as above - - - - - - - - - - 23
23. Corolla tube distinctly curved below then widening towards the mouth, c. 4 mm. diam. at midpoint - - - - - - - - - - 6. *eylesii*
  − Corolla tube more or less straight apart from widening towards the mouth, c. 6−8 mm. diam. at midpoint - - - - - - - 7. *wittei*

1. **Streptocarpus grandis** N.E. Br. in Curtis, Bot. Mag. 131: t. 8042 (1905). Type from S. Africa.
Subsp. septentrionalis Hilliard & Burtt, Streptocarpus 179 & 389 (1971). Type: Zimbabwe, Chimanimani Distr., boundary between Tarka and Glencoe Forest Reserves, Chambuka R., *Goldsmith 3/68* (E, holotype; NU; SRGH).

Unifoliate. Leaf up to c. 185 × 125 mm., rounded at base, crenulate, shortly pilose on both surfaces. Peduncles c. 150−250 mm.; pedicels up to 17 mm.; both pilose, the hairs sometimes gland-tipped. Calyx segments c. 3 mm. long, linear. Corolla 20−30 mm. long, pilose outside, all the hairs acute, or occasionally some gland-tipped; tube 10−20 mm. long, c. 4 mm. across the throat. Filaments arising c. 13 mm. above base of corolla, c. 4 mm. long; anthers 1.5 mm. Ovary 10−15 mm. long, pubescent with semi-patent hairs.

Zimbabwe. E: Chimanimani Mts., fl. & fr. 14.ii.1960, *Goodier 902* (SRGH). **Mozambique. MS:** Chimanimani Mts., Mozambique side of Skeleton Pass, fl. & fr. 11.iv.1967, *Plowes 2619* (SRGH).
Only known from the Chimanimani district of Zimbabwe and adjoining Mozambique. On damp quartzitic rock in stream gullies.

Streptocarpus grandis subsp. *grandis* is found in Natal, where it shows a considerable range of variation (see Hilliard & Burtt, 1971, p.181-183). The disjunct subsp. *septentrionalis* is a smaller plant, less glandular with semi-patent (not patent) hairs on the ovary.

2. **Streptocarpus michelmorei** B.L. Burtt in Kew Bull. **1939**: 72 (1939); in Curtis, Bot. Mag. **166**: t. 54 (1949).—Hilliard & Burtt, Streptocarpus 182 fig. 27D, a & b, pl. 6b (1971). Type: Zimbabwe, Chipinga Distr., Inyamadzi R., near Mt. Selinda, *Michelmore*, cult. in Hort. Bot. Reg. Kew 1938 (K, holotype).

Monocarpic. Leaf solitary, up to 350 × 300 mm., usually more or less cordate at the base, shortly pilose on both surfaces, subsessile or with a short thick stalk. Inflorescences arising from top of stalk and base of midrib, one to several, floriferous with several flowers open at a time; peduncles shortly but densely pubescent; pedicels up to 25 mm.; bracts small, linear. Calyx divided to the base into 5 linear-lanceolate segments 2−5 mm. long. Corolla 30−50 mm.; medium violet on outside of tube and lobes and on inside of lobes, with a deep violet patch on the palate and behind that, typically, a well marked yellow bar; tube 25−30 × 4 mm., subcylindrical, somewhat deepened on the lower side above the middle, mouth wide open, almost circular; limb almost regular, upper lobes c. 4 × 7 mm.; inferior lip c. 11 mm. long, lateral lobes c. 4 × 7 mm., anterior c. 6 × 7mm., more or less porrect. Stamens arising in middle third of tube; filaments 5 mm., thick, curved, glabrous; anther lobes 1.5 mm. Staminodes small. Ovary c. 20 mm. long, densely pubescent with short spreading hairs; style 10 mm., pubescent; stigma stomatomorphic. Capsule very slender, pubescent, up to 130 × 1 mm. Seeds 0.5−6 mm. long, reticulate.

Zimbabwe. E: Chimanimani Distr., Ngorima Reserve, 4000 m., fr. 11.i.1961, *Plowes* 2131 (K; SRGH). **Mozambique.** MS: tributary of Busi R., on rd. to Gogoi, c. 2500 m., fr. 1.i.1962, *Goldsmith 24/62* (E; K; LISC; SRGH).

Populations at Bridal Veil Falls near Chimanimani may well represent hybrids between *S. michelmorei* and *S. eylesii*, whose distributional ranges probably meet about here. These plants tend to have the deeper colouring and slender fruits of *S. michelmorei*, but the distinctly curved corolla tube and larger median lobe strongly suggest the influence of *S. eylesii*. There is considerable variation within these populations. Similar plants from elsewhere (*Cookson 1363* from Vumba Mts. (NU), *Chase 8234* from Ngorima Reserve (SRGH)) need further field studies.
Plants closely allied to *S. michelmorei* are known from Malawi (2A) and from Zambia (2B). Unfortunately the Malawi plant is known only from a single locality, while we have seen only dried material from Zambia and it is very desirable that living material be examined before taking a decision whether the plant can be included in *S. michelmorei*.

2A. Streptocarpus aff. **michelmorei** Hilliard & Burtt, Streptocarpus 183, fig. 27 Ea, b, pl. 6c.

Unifoliate, (?) sometimes plurifoliate at least in cultivation. Leaf thicker than usual in *S. michelmorei*, dark green, the veins very prominent below, tinged red, tending not to lie flat in cultivation. Calyx segments 4−6 mm. long. Corolla 45 mm., deep violet outside, except on belly of tube and on limb, paler or with a yellow bar in the throat; tube 33 mm. long and 4 mm. in diam., subcylindrical, slightly curved and somewhat swollen on the lower side near the mouth; limb somewhat oblique, upper lobes 5 × 6 mm., erect-recurved; lower lip more or less porrect, 10 mm. long, 15−20 mm. across, lateral lobes 6 × 8 mm., median 6 × 7 mm. Stamens arising in upper third of corolla tube, purple, thickened, with a few glands at top; anther-lobe 1 mm. long, purple outside. Ovary 18 mm. long, dull green flushed reddish with short spreading eglandular hairs; style 12 mm. long, white flushed reddish at base and on bases of hairs. Fruit slender, c. 80 mm. long.

Malawi. N: N. Vipya, Chimaliro Hill, *Greig* cult. R.B.G. Edinburgh C. 5366 (E; NU); ibidem, fl. 16.ii.1965, *Chapman 232B* (E).
Only known from this locality. On rocks just outside and on rocks and tree trunks inside the forest; c. 2010 m.

2B. Streptocarpus aff. **michelmorei** Hilliard & Burtt, Streptocarpus 186 (1971).

Unifoliate, leaf at flowering 180 × 150 mm., pilose-pubescent on both surfaces, cordate at base. Inflorescences branched near base, spreading-pubescent throughout with a few glandular hairs interspersed. Calyx segments only 2−3 mm. Corolla c. 40 mm., tube straight, c. 30 mm., only slightly widened upwards, limb almost equally 5-lobed, lobes c. 8 × 7 mm. Stamens arising c. 18 mm. from base of the corolla; filaments 8 mm., glandular towards top. Young fruit 70 mm. with 10 mm. style still persisting; this soon breaks off to give a sharp-pointed fruit, c. 1.5 mm. in diam.

Zambia. W: Solwezi, Kabompo Gorge, 15.v.1965, *Mutimushi 3335* (K; NDO; SRGH). C: Serenje Distr., Kundalila Falls, c. 53 km. ENE. of Serenje, 1400 m., fl. i.1969, *Williamson 1727* (K; SRGH).

3. **Streptocarpus solenanthus** Mansfeld in Notizbl. Bot. Gart. Mus. Berl.-Dahlem 12: 96 (1934).—Schlieben in Gartenflora 85: 274 (1936).—Burtt in Notes Roy. Bot. Gard. Edinb. 22: 572 (1958).—Hilliard & Burtt, Streptocarpus 185, fig. 27, Fa & b, pl. 6,d (1971). Type from Tanzania.

Monocarpic. Leaf solitary, ovate or oblong, usually c. $70-150 \times 40-120$ mm. but up to 350×170 mm., green on both surfaces, shortly pilose. Inflorescence with flowers tending to be massed at one level, eglandular pilose-pubescent, the hairs in the lower part longer and deflexed; peduncles $1-4$, up to 150 mm; pedicels up to 20 mm; bracts linear, c.5 mm. Calyx divided almost to the base into 5 linear-lanceolate segments $2-3$ mm. long, shortly pubescent. Corolla light violet; tube $18-26$ mm. long, cylindric, $3-5$ mm in diam., $22-36$ mm. long, pubescent outside, sometimes wholly white, sometimes the tube and outside of limb whitish, the inner surface of limb light to medium violet, often with slightly darker patches in the throat, occasionally the whole corolla slightly curved towards the top and scarcely widened to the mouth; limb regular, lobes small in relation to tube, more or less spreading, upper two divergent, $2-4 \times 2-5$ mm., lower lip $5-8$ mm. long, lobes $3-6 \times 2.5-6$ mm., all rounded. Stamens arising in upper third of corolla tube; filaments 4 mm. long, somewhat thickened about the middle, glabrous; anther-lobe c. 0.75 mm., cream tinged mauve. Ovary c. 15 mm. long, green, densely covered with short spreading eglandular hairs, passing into the white $6-7.5$ mm. long pubescent style; stigma stomatomorphic, strongly papillose. Fruit slender, shortly pubescent, $50-80$ mm. long. Seeds $0.4-0.5$ mm. long, reticulate.

Zambia. N: Nyika, Chowo forest, i.1967, *Hilliard & Burtt* 4381, fl. R.G.B. Edinb. vi.1967, C.5353 (E,NU). **Zimbabwe**. E: Mutare Distr., Mtasa North reserve, Mpanda R., tributary to Hondi R., 11.ii.1949, *Chase* 1157 (K; LISC; SRGH). **Malawi**. N: Chitipa Distr., Misuku Hills, Mugesse Forest Reserve, 16.iii.1977, *Grosvenor & Renz* 1240 (E; K; SRGH).
Also in Tanzania. Epiphytic in forest or growing on exposed rock outcrops; $1350-2200$ m.

A few specimens from the Mutare district of Zimbabwe suggest, by the more flared corolla mouth, that hybridisation with *S. eylesii* may have taken place: such are *Chase* 5193 & 7945 (SRGH), Burtt 3072(E). One specimen from the Nyika Plateau in northern Malawi, *Simon, Williamson & Ball* 1746 (SRGH), also has a more flared corolla with large median lobe 1 cm. long: this may result from hybridisation with *S. wittei*, common in that neighbourhood.
The form with pure white flowers is probably restricted to the Mulange Distr., and the Uluguru Mts. in central Tanzania. Plants from further south always have at least a tinge of mauve in the corolla.

4. **Streptocarpus goetzei** Engl., Bot. Jahrb. 30: 406 (1901).— Burtt in Kew Bull. 1939: 70 (1939); in Mem. N.Y. Bot. Gard. 9: 17 (1954).—Hilliard & Burtt, Streptocarpus 186, fig. 27, Ha & b, pl. 7, c (1971). Type from Tanzania.
Streptocarpus mahonii J. D. Hook. in Curtis, Bot. Mag. t. 7857 (1902).—Baker & Clarke in F.T.A. 4, 2: 505 (1906). Type: Cult. in Hort. Bot. Reg. Kew, 1900; seed sent from Africa (probably from near Zomba, Malawi) by *J. Mahon* (K).
Streptocarpus rungwensis Engl. (incl. var. *latifolius*), Bot. Jahrb. 57: 218 (1921). Type from Tanzania.
Streptocarpus lujai De Wild. in Rev. Zool. Afr. 5 Suppl. Bot.: 272 (1932). Type: Mozambique, Marrumbala, *Luja* 38, (BR, holotype).
Streptocarpus tubiflos sensu Baker & Clarke in F.T.A. 4, 2: 506 (1906) non C.B.Cl.
Streptocarpus breviflos sensu Baker & Clarke in F.T.A. 4, 2: 506 (1906) non C.B.Cl.

Monocarpic. Leaf solitary, sessile, commonly about 150×120 mm. but can be much smaller and also much larger and can be proportionately much narrower (250×65 mm.), more or less densely pubescent on both surfaces, the indumentum usually erect in living plant. Inflorescences 1 to about 6, rather densely clad with short spreading eglandular hairs, occasionally with a few glandular ones on the pedicels; peduncle of the primary one often c. 100 mm. long. Bracts linear, $2-3$ mm., pubescent. Calyx segments varying from (2) $3-5$ (6) mm., linear-lanceolate, pubescent. Corolla (20) $25-35$ mm. long, medium violet with 2 darker patches at sides of mouth, thinly pubescent with short erect hairs; tube $12-22$ mm. long, $2-3.5$ mm. in diam. and curved about mid-way, mouth compressed between the upper lobes so that the opening is shaped like an inverted V, 6−8 mm. wide between lateral sinuses; upper lobes erect, c. 5 × 6 mm., divergent or somewhat divaricate; lower lip 10−14 mm. long, 15−20 mm. broad, more or less porrect, lateral lobes c. 6 × 7 mm., rounded, median lobe 6−7 × 6−9 mm. Stamens arising from a little above half way to a little above two thirds up the corolla tube; filaments 2−2.5 mm., glandular towards top; anther lobe 0.75−1 mm. Ovary 7−9 mm. long, densely clad with short spreading eglandular hairs, passing into the 1.5−2 mm. long style; stigma capitate, stomatomorphic. Capsules slender, 40−60 mm. × 1.5 mm., shortly pubescent. Seeds 0.5−0.6 mm. long, reticulate.

Malawi. N: Rumphi Distr., NE. of Livingstone, Manchewe Falls, 1000 m., 17.iii.1977, *Grosvenor & Renz* 1252 (SRGH). S: Cholo Mt., 1220−1460 m., 16.ii.1970, *Brummitt & Banda* 8598 (K; MAL; SRGH). Mozambique. Z: Serra da Gúruè, 1100 m., 21.ii.1966 *Torre & Correia* 14777 (K; LISC; SRGH). N: Ribáuè, Mepalué, c. 1350 m., fl. 25.i.1964, *Torre & Paiva* 10242 (LISC).

Also in Tanzania. Damp shady banks and rock crevices or overhangs, or epiphytic on mossy trunks.

5. **Streptocarpus confusus** Hilliard in Journ. S. Afr. Bot. 32: 115 (1966). Type from S. Africa.

Subsp. **lebomboensis** Hilliard & Burtt in Notes Roy. Bot. Gard. Edinb. 28: 209 (1968), Streptocarpus 304 fig. 44M & 46 (map) (1971). Type from S. Africa.

Monocarpic. Leaf solitary, up to 220 × 150 mm., margin crenate, base cordate, dark green above, usually purplish-red below, pilose, sometimes glandular. Inflorescences: the first often from the apex of the hypocotyl, thereafter from the upper surface of the leaf at the base of the midrib, c. 16-flowered, few open together. Peduncles up to 30 cm. long; pedicels c. 8−18 mm. long, both pilose with glandular and eglandular hairs. Calyx divided to the base into 5 segments, 3−5 × 1−1.5 mm. Corolla 35−55 mm. long, pilose with glandular and eglandular hairs outside, minutely glandular-pubescent inside near the mouth; tube subcylindrical, abruptly deflexed then directed forwards, laterally compressed in the mouth, 25−30 mm. long, up to 5 mm. broad, pale violet to whitish; limb oblique, more conspicuous than the tube, pale to medium violet, throat white or pale yellow, lateral sinuses very narrow, upper lip of 2 divaricate lobes, lower lip 10−20 mm. long, 12−30 mm. broad, all lobes oblong to suborbicular, 7−12 × 5−11 mm. Stamens arising about halfway up corolla tube, filaments c. 4.5−6 mm. long, sparsely glandular; anther lobes 1 mm. long, white. Ovary 10−20 mm. long, appressed-pubescent; style terete, c. 5 mm. long; stigma bilobed with the lobes held laterally. Capsule 55−75 mm. long, 2 mm. in diam. Seeds 0.6−0.8 mm. long, reticulate.

Mozambique. M: Namaacha, Changalane, Estatuine, fl. & fr. 10.v.1969, *Balsinhas* 1483 (LISC).

Streptocarpus confusus subsp. *lembomboensis* is restricted to the Lebombo Mts. from the Mkuzi R. gorge and Ubombo in Natal through Swaziland to Namaacha in Mozambique, the only known locality within the Flora Zambesiaca Area. *S. confusus* subsp. *confusus* ranges from the Natal Midlands to the north-eastern Transvaal and is distinguished by its subcapitate stigma and smaller flowers. *S. confusus* is closely allied to the S. African *S. haygarthii* C.B.Cl.

6. **Streptocarpus eylesii** S. Moore in Journ. Bot. 57: 245 (1919).—Norlindh in Bot. Notis. 1948: 36 (1948).—Schelpe in Fl. Pl. Afr. 33: t. 1304 (1959).—Burtt in Curtis's Bot. Mag. 177: t. 541 (1969).—Hilliard & Burtt, Streptocarpus 193 (1971). Type: Zimbabwe, Matopo Hills, in wet cavities under shadow of granite rocks, *Eyles* 1097 (BM, holotype; PRE; SRGH; Z).

Unifoliate monocarpic herb, more rarely plurifoliate and perennial. Leaf up to 30 × 20 cm., pilose-pubescent on both surfaces, irregularly crenate-dentate on the margin, subsessile or with a short stalk up to 20 mm. on which the hairs point downwards. Inflorescences arising from top of stalk and base of midrib, c. 30 cm. high. Peduncles 12−25 cm., pilose-pubescent; branches of inflorescence either pilose-pubescent or with greater or lesser admixture of glandular hairs. Bracts usually linear, c. 10 mm. Pedicels c. 25−35 mm. Calyx divided almost to base into 5 linear segments 3—9 mm. long, pilose-pubescent with or without some glandular hairs. Corolla 37—65 mm. long, pilose or glandular-pilose outside, often pale or whitish outside, pale to medium violet on the limb, usually with darker patches at the mouth, rarely with a pale yellow patch in the white throat; tube 25—40 mm., 3—4 mm. in diam. at mid-point; upper lip of 2 rounded lobes c. 7—10 × 10—11 mm., lower lip 20—30 mm. long, lateral lobes 9—15 × 9—18 mm., median 10—14 × 10—18 mm. Stamens arising in upper third of corolla tube; filaments 5—8 mm. long, thickened above base, generally with stalked glands in the upper part; anther lobes 1.75—2 mm. long. Ovary 6—24 mm. long, clad with dense spreading glandular and eglandular hairs, passing into style of 2—10 mm., glandular-pubescent; stigma stomatomorphic, 2—3 mm. broad. Capsule c. 40—75 × 2—3 mm., glandular-pubescent, glabrescent after ripening. Seeds 0.6—0.9 mm. long, reticulate.

A very variable species in itself and part of the still wider *S.* agg. *monophyllus.* Three subspecies are recognised.

1. Style elongate, reaching to mouth of corolla tube; plants always unifoliate
 - - - - - - - - - - - - - - subsp. *eylesii*
— Style short, not reaching beyond the flexure of the corolla tube; plants often plurifoliate, sometimes unifoliate - - - - - - - - - 2
2. Corolla limb mauve to purple usually without a yellow mark in the throat; amongst rocks on grassy hillsides - - - - - - - subsp. *brevistylus*
— Corolla limb white with well-marked yellow patch in throat; *Brachystegia* woodland
 - - - - - - - - - - - - - subsp. *silvicola*

6a. Subsp. eylesii. Hilliard & Burtt, Streptocarpus 194 (1971). Type as for the species.

Unifoliate. Calyx segments 5.5—7 mm. long. Corolla 40—60 mm. long; tube 30—40 mm. long. Filaments 5—8 mm. long. Ovary 12—24 mm. long; style 6—10 mm. long. Capsule 40—75 mm. long.

Zambia. W: Solwezi-Mwinilunga Rd., Kabompo R., 1400 m., fl. & fr. 15.ii.1975, *Hooper & Townsend* 57 (K). **Zimbabwe.** N: Goromonzi Distr., Damusi, Chindamora Forest Reserve, fl. & fr. 26.ii.1967, *Grosvenor* 302 (E; NU; SRGH). W: Matopo Hills, fl. xi.1902, *Eyles* 1097 (BM; SRGH; Z). E: Mutare, Vumba Mts., fl. 1.i.1966, *Plowes* 2583 (E; NU; SRGH). S: Belingwe Distr., Mt. Buhwa, 27.iv.1973, *Pope* 954 (K; SRGH). **Malawi.** C: Dedza Mt., fl. & fr. 4.i.1967, *Hilliard & Burtt* 4164 (E; MAL; NU). **Mozambique.** T: Angonia, Zóbuè para Matengobalama, 11.i.1966, *Correia* 402 (E; LISC). MS: Mt. Gorongosa, 25.vii.1970, *Müller & Gordon* 1461 (SRGH); ibidem, c. 1000 m., 6.v.1964, *Torre & Paiva* 12263 (LISC).
Usually among rocks and grass on open hillsides. For notes on possible hybrids see under *S. michelmorei* and *S. solenanthus.*

6b. Subsp. brevistylus Hilliard & Burtt in Notes Roy. Bot. Gard. Edinb. **28**: 210 (1968); Streptocarpus 194 (1971). Type: Malawi, S. Vipya, Wozi Hill near Chikangawa, *Hilliard & Burtt* 4204 (E, holotype; MAL; NU, isotypes).

Plurifoliate or sometimes unifoliate. Calyx segments 6—9 mm. long. Corolla 37—60 mm. long; tube 22—35 mm. long. Filaments 5 mm. long. Ovary 6—10 mm. long; style 2—3 mm. long. Capsule 40 mm. long.

Zambia. N: Mbala Distr., Kambole escarpment, 2.i.1955, *Richards* 3871 (K). **Malawi.** N: Mzimba Distr., S. Vipya, Champira, Lwanyati Peak, 11.i.1975, *Pawek* 8908 (K; MAL; SRGH). S: Mt. Malosa, 3.i.1967, *Hilliard & Burtt* 4155 (E). **Mozambique.** MS: Zembe Mt., c. 28 km. S. of Vila Pery, 666 m., *Plowes* cult. R.B.G. Edinb., C8087 (E). Z: Gúruè, serra de Gúruè subida pela picada Chã Mozambique, near the source of R. Malema, c. 1750 m., 4.i.1968, *Torre & Correia* 16898 (LISC).

6c. Subsp. silvicola Hilliard & Burtt in Notes Roy. Bot. Gard. Edinb. **43**: 231 (1986). TAB. 8. Type: Malawi, Lilongwe Distr., c. 56 km. SW. of Lilongwe, Dzalanyama Forest Reserve, Chaulongwe Falls, *Hilliard & Burtt* 4486 (E, holotype; NU).

Unifoliate. Calyx segments 5—8 mm. long. Corolla white with yellow patch on floor of throat, c. 55 mm. long; tube c. 35 mm. long. Ovary c. 6 mm. long; style 2 mm. long; stigma 2 mm. in diam.

Malawi. C: Dzalanyama Forest Reserve, Chaulongwe Falls, fl. 28.iii.1970, *Brummitt* 9484 (K). **S:** Mangochi Distr., Mt. Uzuzu, fl. 21.i.1971, *Hilliard & Burtt* 6301 (E; NU); Phirilongwe, fl. & fr. 17.iii.1985, *Johnston-Stevard* 410 (E).
Only known from three localities. Among rocks, or epiphytic on fallen trunks, in *Brachystegia* woodland; c. 1000−1350 m.

This subspecies merits recognition because its constantly white flowers are associated with a distinctive habitat, *Brachystegia* woodland, where subsp. *brevistylus* has not been found. Subsp. *silvicola* is now known from three well separated localities and should be sought in intervening areas. When first discovered it was treated as a white variant of subsp. *brevistylus* (Hilliard & Burtt 1971, pp. 196 & 385).

7. Streptocarpus wittei De Wild. in Rev. Zool. Afr. 5, Suppl. Bot.: 90 (1932).−Burtt in Notes Roy. Bot. Gard. Edinb. 22: 573 (1958).−Hilliard & Burtt, Streptocarpus 196, fig. 30A a & b, pl. 8 f (1971). Type from Zaire.
    Streptocarpus katangensis sensu De Wild. in Ann. Mus. Cong. Ser. 4, 1: 127 (1903).−vix De Wild. & Th. Dur.

Unifoliate monocarpic herb. Leaf broadly ovate, up to 25 × 28 cm., base cordate; upper surface shortly but rather densely pilose, the lower surface densely pilose on the nerves, more sparingly so in between; margin shallowly crenate. Inflorescences 1−5, many-flowered, several flowers open simultaneously on each. Peduncles 10−20 cm. long, densely pilose-pubescent with eglandular and (especially towards top) glandular hairs. Pedicels densely glandular-pilose, 30−45 mm. long. Bracts usually linear, c. 8 × 1.5 (3) mm. long, densely pilose. Calyx segments 5−10 × 1.5 mm., linear, obtuse, densely pilose-pubescent, many of the hairs often glandular. Corolla 50−60 mm. long; tube 30−40 × 6−7 mm., almost straight or slightly deflexed and widened above the middle, most of the short spreading hairs on the outside glandular; upper lip of 2 rounded lobes 8−13 × 7−10 mm., lower lip 17 mm. long, median lobe 11 × 10 mm., laterals 9 × 9. Stamens arising in upper third of corolla tube; filaments 6−8 mm., glandular towards top; anther lobes 2 mm. Ovary 20 mm. long, densely short spreading pubescent, most of the hairs eglandular, passing into style 10 mm. long and becoming more glandular upwards; stigma stomatomorphic. Capsule 50−90 (usually c. 80) mm. long. Seeds 0.4−0.6 mm. long, reticulate.

Zambia. E: Nyika, fl. i.1961, *Benson* NR 474 (K). **Malawi. N:** Nyika, Chowo Rocks, 6.ii.1968, *Williamson, Ball & Simon* 1667 (SRGH). **S:** Mt. Mulanje, Luchenya Plateau, seed coll. *Richards* 16530, cult. R.B.G. Edinb. C4752 (E).
Also in Zaire. Among rock outcrops; c. 2100 m.

8. Streptocarpus arcuatus Hilliard & Burtt in Notes Roy. Bot. Gard. Edinb. 43: 229 (1986). Type: Malawi, Blantyre Distr., Mpingwe Hill near Limbe, 29.i.1967, *Hilliard & Burtt* 4137 (E, holotype).

Unifoliate. Leaf of flowering plant always withered at apex where the terminal part has abscinded during the unfavourable season, up to 30 cm. long and 20 cm. broad, pilose on both surfaces. Peduncle up to 15 cm. long, with spreading glandular and eglandular hairs. Calyx segments 5.5 mm. long. Corolla (to tip of extended lower lobe) 7−8 cm. long, arcuate, inflated above the base on the upper side, maximum diam. of tube (as pressed) 13 mm.; corolla tube white, lobes white outside but with the medium violet of the inner surface showing through, floor of tube with deeper violet blotch at base of inferior lip, two yellow spots usually present below sinuses of inferior lip. Stamens arising 4 cm. above the base of tube; filaments 15 mm. long. Gynoecium 6 cm. long.

Malawi. S: Ndirande Mt. near Limbe, *Cram* cult. in R.B.G. Edinb. C 5398 (E).
Known only from Mpingwe and Ndirande, near Limbe. Amongst rocks on grassy slope.

Originally (Hilliard & Burtt 1971, p.191) placed as "sp. aff. *S. wittei*"; it has become desirable to give this plant an independent name.

9. Streptocarpus myoporoides Hilliard & Burtt in Notes Roy. Bot. Gard. Edinb. **28:** 213 (1968); Streptocarpus 213, fig. 32D a & b, pl. 1 (1971). Type: Mozambique, Ribáuè,

Tab. 8. STREPTOCARPUS EYLESII Subsp. SILVICOLA. 1, habit (×⅔); 2, calyx and gynoecium (×1); 3, corolla opened out (×1); 4, stamens (×4); 5, gynoecium and disk (×2), 1−5 from *Hilliard & Burtt* 4486. 6, capsule (×1), from *Hilliard & Burtt* 6301.

Serra de Ribáuè (Mepalue), c. 1500 m., *Torre & Paiva* 10302 (E; LISC, holotype; SRGH).

Apparently perennial from a rhizome. Two possible growth-patterns have been observed on herbarium material. (A) the rhizome produces several annual shoot systems, each shoot consisting of an erect stem up to 30 × 8 mm., rooting at the base and probably at least partly subterranean, with a pair of large cauline leaves at the apex and terminal inflorescences; (B) two large basal leaves are produced (not exactly opposite each another), the lower bears small vegetative leaves at base of petiole, the other gives rise to the inflorescence axis at the base of its petiole and this bears a pair of small cauline leaves c. 5 cm. above the base. Leaves (basal or large cauline) elliptic, c. 15−25 × 5−10 cm., apex obtuse, base cuneate, decurrent on the petiole (c.30 mm.), margin coarsely serrate-dentate, softly pilose above, the hairs more or less confined to the nerves below, nerves and stalk reddish-brown; small cauline leaves somewhat unequal, 30−50 × 12−15 mm. Peduncles terminal, c. 45−60 × 5 mm., densely pilose with whitish hairs, and with a pair of reduced leaves c. 15 × 5 mm. well below the inflorescence. Inflorescence capitate (but the material is very young) subtended by four ovate bracts, c. 9 × 8 mm., with serrate margins, pilose, apparently deep red. Pedicels 4−5 mm. long, stout, sparingly pilose, each subtended by a bract, c. 10 × 2 mm., closely resembling a calyx segment. Calyx divided nearly to the base into five lanceolate-falcate segments, 3 upper 11 × 3 mm., erect and with curved tips, 2 lower 13 × 3 mm., directed forwards, all with 3−5 conspicuous parallel nerves, small dark glands inside, coarse appressed hairs outside, margins ciliate. Corolla c. 25 mm. long, "vermilion", with coarse appressed hairs outside, glabrous inside except for gland-tipped hairs around the mouth and on the lower lip; tube broadly funnel-shaped, held obliquely erect from the contracted base, c. 23 mm. from the base to the sinus between the two posticous lobes, c. 6 mm. across mouth, hooded owing to the two lateral lobes ascending to join the two posticous lobes and so forming the upper lip, posticous lobes suborbicular, c. 3 × 3 mm., the two laterals elliptic and slightly larger, lower lip (anticous lobe) elliptic, c. 11 × 5 mm., deflexed. Stamens arising in upper third of corolla tube and protruding from corolla mouth; filaments c. 8 mm. long, slightly thickened about the middle and with a few gland-tipped hairs; anther lobes c. 2 mm. long, lateral staminodes 1 mm. long. Ovary c. 5 mm. long, appressed- pubescent; style c. 11 mm. long, more or less terete, curved up under the corolla hood and slightly exserted, pubescent at base, becoming glabrous upwards; stigma terminal, barely broader than the style, a grooved papillose dome rimmed by stylar tissue. Capsule bearing elongated (up to 25 mm.) style until nearly ripe, c. 30 × 1.5 mm., pilose. Seeds 0.4−0.5 mm. long, verruculose.

Mozambique. N: Serra Chinga, *Macedo* 3167 (LMA).
Only collected twice. Further material, and seed for cultivation, is needed for the growth pattern of this species to be elucidated.

10. **Streptocarpus milanjianus** Hilliard & Burtt, Streptocarpus 385 and 390 (1971). Type: Malawi, Mt. Mulanje, 1890 m., rocks by Thuchila R., at crossing of Thuchila-Chambe path, 3.ii.1971, *Hilliard & Burtt* 6385 (E, holotype; MAL; NU).

Perennial, rosulate, rhizomatous herb. Leaves up to c. 10, 2-ranked at the apex of the rhizome, the flowering ones with conspicuous petiolodes; petiolode curved, up to about 10 cm. long, reddish-purple, pilose with deflexed to spreading reddish-purple hairs; lamina elliptic, up to 33 × 13 cm., often much smaller, base cuneate, not or scarcely decurrent on the short petiole, silky appressed-pilose above, very densely short-pilose on the nerves below and sparingly so between the nerves, margin irregular-serrate, lateral nerves numerous, spaced at 5−10 mm., reddish-purple below. Inflorescences 1−3, arising a little below the base of the lamina, up to 20-flowered, the cymes open and spreading. Peduncles up to 20 cm., reddish-purple, with spreading reddish-purple eglandular hairs in the lower part and some glandular ones intermixed above; pedicels up to 20 mm., with most of the hairs glandular. Bracts linear, up to 7 mm. long, pilose. Calyx divided to the base into 5 lanceolate segments. Corolla c. 15−22 mm. long, glandular-pubescent outside, bearded with long unicellular hairs within on the roof of the tube, white flushed dull purple below on the outside of the tube, with three broad reddish-purple bands inside on the floor, these bands sometimes bifurcate or each split into 2 to 3 narrower stripes; tube 8−12 mm. long, c. 6−8 mm. across the mouth, slightly swollen on the upper side at the base, somewhat deflexed and ventricose on the lower side towards the mouth; upper lobes 4−5 × 3−4 mm.; lower lip c. 7−10 mm. long, projecting forwards, lobes 4−6 × 4−6 mm. Stamens arising in lower third of tube; filaments 2.5−3.5 mm. long, swollen above the middle, purple-spotted; anthers more or less triangular in outline, lobes 1.5 mm. long, scarcely divergent; staminodes minute. Ovary 5−7 mm. long, pilose with spreading eglandular hairs; style 4−6 mm. long, pubescent; stigma stomatomorphic, papillose. Capsule (old) up to 26 mm. long, terminating in the persistent style. Seeds 0.7 mm. long, verruculose.

Malawi. S: Mt. Mulanje, between Madzeka hut and Little Ruo Falls, 12.i.1971, *Hilliard & Burtt* 6327 (E; MAL; NU).
Only known from Mt. Mulanje. On large shady moss-covered rocks by or in streams; c. 1670−1950 m.

11. **Streptocarpus dolichanthus** Hilliard & Burtt in Notes Roy. Bot. Gard. Edinb. 43: 230 (1986). Type: Malawi, Mulanje Mt., Ruo Gorge, c. 1700 m., fl. 17.v.1983, *Johnston-Stewart* in la Croix 486 (E, holotype).

Perennial herb with horizontal rhizome. Leaves about 5, biseriate near the apex of the rhizome; petiolode c.4 cm. long at flowering, petiole c. 5 cm., both pubescent; lamina elongate-oblong or narrowly elliptic, 22−27 × 4−6 cm., with short appressed sericeous hairs on upper surface and veins below, subglabrous between the veins below, lateral veins numerous, c. 5−10 mm. apart, margin irregularly and closely denticulate. Inflorescences 2−14-flowered; peduncle up to 11 cm. long, appressed-pubescent; primary pedicels 1.5−2 cm. long, paired. Bracts (primary) linear, 5 mm. long. Calyx divided nearly to the base into 5 linear segments 7 mm. long. Corolla tube narrowly cylindrical in the lower 5 cm., funnel-shaped in upper 2.5 cm., sparsely spreading-pubescent outside; upper lobes 7 × 7 mm., laterals 10 × 7 mm., median 12 × 8 mm., all obtuse. Stamens with filaments 8 mm. long, glabrous, arising 5 cm. above the base of the corolla; anthers 3 mm. long, 3 mm. broad across the divergent thecae at the base, connective dorsally thickened. Ovary c. 2.5 cm. long, densely appressed-pubescent, gradually narrowed into the 4.5 cm. long style; stigma horizontally bilobed, the lower lobe the larger. Capsule 8 cm. long, tipped by the 2−3 cm. long persistent part of the style.

Malawi. S: Mulanje Mt., Ruo Gorge, fl. 10.iv.1983, *Jenkins* s.n. (E).
Known only from Mt. Mulanje. On shady mossy rocks by or in streams; 1200−1700 m.

The late flowering season of this species (April-June) is noteworthy.

12. **Streptocarpus hirtinervis** C.B.Cl. in F.C. 4, 2: 446 (1904) quoad descript.—Baker & Clarke in F.T.A. 4, 2: 507 (1906).—Burtt in Kew Bull. 1939: 71 (1939); in Mem. N.Y. Bot. Gard. 9: 18 (1954); in Notes Roy. Bot. Gard. Edinb. 22: 574 (1958).—Hilliard & Burtt, Streptocarpus 217 fig. 37B a & b, pl. 9b (1971). TAB. 9. Type: Malawi, Mt. Mulanje, 2100−2700 m., xii.1900, *Purves* 91 (K, holotype).

Perennial, rosulate herb with a horizontal rhizome. Leaves up to about 10 in a tuft at apex of rhizome, only the flowering ones with conspicuous petiolodes up to 60 mm. long, silky-pilose, the hairs often reddish-violet; lamina oblong, up to 20 × 10 cm., often much smaller, base cuneate, shortly decurrent on very short petiole, both surfaces pilose with silky-silvery appressed hairs, midrib and lateral nerves raised below, the lateral nerves forking at their apices and closely outlining the deeply and coarsely serrate leaf margin; lamina plicate between the lateral nerves. Inflorescences usually 1, occasionally 2, at junction of petiole and petiolode, up to c. 20-flowered. Peduncles up to c. 15 cm. long; pedicels up to c. 12 mm. long, both pilose with glandular and eglandular hairs. Bracts linear-oblong, up to c. 6 mm. long, occasionally foliaceous. Calyx divided

Tab. 9. STREPTOCARPUS HIRTINERVIS. 1, habit (×⅔); 2, flower from above (×1); 3, corolla, ventral view (×1); 4, corolla opened out (×1); 5, calyx and gynoecium (×1); 6, gynoecium and disk (×2); 7, stamens (×2), all from *Hilliard & Burtt* 4605.

to the base into five lanceolate segments 4.5 × 1 mm., pilose with glandular and eglandular hairs. Corolla 25−40 mm. long, glandular-pubescent outside, glabrous inside except for unicellular hairs on the roof in the upper half of the tube, medium violet, sometimes with a magenta-pink tinge, paler inside, with 3−7 deeper violet lines on floor of tube and lower lip; tube c. 15−30 mm. long, 5−8 mm. across mouth, subcylindrical, base oblique, saccate on upper side, sometimes markedly so, slightly constricted about the middle and bent downwards, widening slightly upwards; limb almost straight, upper lip of 2 erect suborbicular lobes 3−6 mm. long, lower lip 8−12 mm. long, lobes subcircular, 5−8 mm. long. Stamens arising in upper third of corolla tube, filaments c. 3.5−5 mm. long, slightly thickened upwards, anther lobes 2 mm. long, staminodes minute. Ovary c. 7−10 mm. long, glandular-pubescent; style c. 12−15 mm. long, sparsely pubescent, slightly compressed dorsoventrally; stigma stomatomorphic, papillose. Capsule 30−55 × 2.5 mm., apiculate from the persistent style base. Seeds 0.7−0.9 mm. long, verruculose.

Malawi. S: Zomba, Malosa Mt., 1650 m. fl. & fr. 3.i.1967, *Hilliard & Burtt* 4057 (E; MAL; NU).
Known only from Mt. Mulanje, Mt. Chiradzulu and Malosa and Chifundi peaks on Zomba Plateau, 1600−2700 m. Rock faces and crevices, bases of sedge tufts.

13. Streptocarpus nimbicola Hilliard & Burtt in Notes Roy. Bot. Gard. Edinb. 28: 214 (1968); Streptocarpus 220, fig. 37 D, frontispiece (1971). Type: Malawi, Mt. Mulanje, Chambe basin, *Hilliard & Burtt* 4507 (E, holotype; NU).

Rhizomatous, rosulate, perennial. Leaves many, oblong, up to 15 × 4 cm., apex withered, base cuneate, both surfaces pilose with silky, silvery appressed hairs, midrib and lateral veins raised below, the lateral veins forking at their apices and closely outlining the deeply and coarsely serrate leaf margin; lamina plicate between the lateral veins. Inflorescences 1−3 immediately below the leaf base, up to c. 20-flowered. Peduncles up to c. 15 cm. long, pilose with spreading hairs. Bracts oblong-elliptic, up to 8 × 2 mm. Pedicels c. 8−30 mm. long, green or reddish-brown, pilose or glandular-pilose or mixed. Calyx divided to the base into 5 linear-lanceolate segments c. 8 × 1 mm., indumentum as on pedicel. Corolla c. 15−30 mm. long, tube whitish often heavily suffused magenta-pink above; limb whitish outside, pale to medium violet or pink inside, white around entire mouth, pilose outside, particularly on tube, with or without gland-tipped hairs, densely glandular inside particularly around mouth; tube c. 5−7 mm. long, subcylindrical, bulbous at base, base c. 3 mm. in diam., bent sharply downwards about the middle, compressed laterally in the throat; throat c. 2 mm. across; limb very oblique, more conspicuous than the tube, upper lip of 2 suborbicular to elliptic-oblong lobes, c. 5−10 × 5−8 mm., lower lip c. 10−20 mm. long, lobes suborbicular to elliptic-oblong, c. 5−12 × 4−12 mm. Stamens arising in upper third of corolla tube; filaments c. 2 mm. long, tinged pink; anther lobes 1.5 mm. long, blue-black at maturity; staminodes minute. Ovary 3−5 mm. long, pubescent with spreading hairs, gland-tipped or not; style c. 2 mm. long, terete, strongly curved; stigma capitate, epapillose, jelly-like. capsule 30−40 mm. long, occasionally up to 10 mm. shorter or longer. Seeds 0.5 mm. long, verruculose.

Malawi. S: Mt. Mulanje, Fort Lister to Sombani, 31.xii.1970, *Hilliard & Burtt* 6071 (E; NU).
Only known from Mt. Mulanje. In crevices of cliffs and rock masses, on moss or lichen-covered rock sheets and in the fibrous bases of tufts of the sedge *Coleochloa setifera*; 1500 to at least 2150 m.

14. Streptocarpus brachynema Hilliard & Burtt, Streptocarpus 222 and 387, fig. 37 S a & b, pl. 10 d (1971). Type: Mozambique, Manica e Sofala, Gorongoza massif, c. 1590 m., *Wright* 546, cult. in R.B.G. Edinb. C6073 (E, holotype).

Monocarpic. Leaf solitary, about 15 × 15 cm. at flowering, with erect hairs on both surfaces, cordate at the base, crenulate. Inflorescences 3−4 at base of lamina, many-flowered; peduncle 10 cm., spreading-pubescent, a few of the hairs glandular; pedicels c. 15 mm., glandular-pubescent; bracts usually linear,

green, c. 8 × 1 mm. (occasionally broader and more leafy in the cultivated plant).
Calyx pubescent, divided to the base into 5 linear segments 4 mm. long. Corolla
15−17 mm. long; tube 7 mm. long, c. 5 mm. diam., slightly swollen on the upper
side at the base and ventricose below towards the mouth, glandular-pubescent
and pale violet outside glabrous and white inside in the lower half, bearded with
pointed unicellular hairs on the roof and sides in the upper half, with a median
yellow bar on the floor and an irregular deep violet (sometimes broken) bar on
each side of it; upper lobes 4 × 4 mm.; lower lip 10 mm. long, the palate white;
median and lateral lobes 5 × 5 mm.; all the lobes pale violet on both sides, almost
glabrous outside, densely papillose all over the surface inside. Stamens arising
in lower third of the tube (about 1.5 mm. above base); filaments 2.5 mm.,
glabrous, rather thick; anthers cohering face to face, buff-coloured, lobes 0.75
mm. long, disk annular. Ovary 5 mm., cylindrical, densely aprressed-pubescent;
style 5 mm., glabrous in the upper part, stigma with a central horizontal groove
and thus almost two-lipped but also very slightly folded forwards about the
vertical axis. Capsule 30−40 × 1 mm., glabrescent. Seeds 0.5 mm. long,
reticulate.

Mozambique. MS: Gorongoza Mt., fl. & fr. 15.iii.1969, *Tinley* s.n. (NU).
Only known from Mt. Gorongoza. On rocks or tree trunks in forest or in *Philippia* zone
above forest; 1590−1800 m.

15. **Streptocarpus umtaliensis** B.L. Burtt in Kew Bull. **1936**: 492 (1936).—Hilliard & Burtt,
    Streptocarpus 223, fig. 37 M a & b, pl. 10 b (1971). Type: Zimbabwe, Mutare, 1560
    m., *Eyles* 4475 (K, holotype; SRGH).

Monocarpic. Leaf solitary, oblong, up to c. 20 × 10 cm., base cuneate to
subcordate, margin finely dentate-serrate, both surfaces appressed-pubescent.
Inflorescences several from base of midrib, many-flowered. Peduncles up to c.
25 cm. long; pedicels 5−20 mm. long, both pubescent. Bracts minute. Calyx
divided to the base into 5 lanceolate segments c. 2.5 × 1 mm., pubescent. Corolla
c. 11−13 mm. long, pure white or very rarely violet-blue, pubescent outside,
glabrous inside except for minute unicellular hairs on roof and floor of tube
and inside of lobes; tube 7−9 mm. long, 3−4 mm. diam., subcylindrical, directed
downwards then forwards, somewhat ventricose below mouth, base slightly
oblique and slightly saccate above; limb straight, lobes of both lips directed
forwards, all lobes oblong, subequal, c. 2 mm. long. Stamens arising in lower
third of corolla tube, filaments c. 3 mm. long, slightly thickened and glandular
above; anther lobes c. 1 mm. long; staminodes minute. Ovary c. 4 mm. long,
appressed-pubescent; style c. 4.5 mm. long, glabrous above, slightly
dorsoventrally compressed; stigma elliptic, stomatomorphic, papillose. Capsule
about 35 mm. long, 1.5 mm. broad, crowned with the persistent style base. Seeds
0.7−0.9 mm. long, reticulate.

Zimbabwe. E: Vumba Mts., Cloudlands, 27.ii.1949, *Chase* 1374 (K; SRGH). **Mozambique**.
MS: Baruè, Serra de Choa, 28.iii.1966, *Torre & Correia* 15478 (E; LISC).
Epiphytic on tree trunks, occasionally on mossy rocks, in evergreen forest; c. 1400−
1850 m.

16. **Streptocarpus erubescens** Hilliard & Burtt in Notes Roy. Bot. Gard. Edinb. 28: 210
    (1968); Streptocarpus 243, fig. 37, Q a & b (1971). Type: Malawi, Blantyre, Ndirande
    Mt., 1.ii.1967, *Hilliard & Burtt* 4657 (E, holotype; NU).

Monocarpic, all parts pubescent with spreading hairs, a few gland-tipped hairs
on the leaf below. Leaf solitary, up to about 15 × 12 cm., often much smaller,
dark green above, sometimes purple below, margin crenate, base cordate to
cuneate. Inflorescences 2 or 3 from base of midrib, few-flowered. Peduncles up
to c. 10 cm. long. Bracts small, linear, occasionally foliaceous. Pedicels 15−30
mm. long. Calyx divided to the base into 5 linear segments 5−9 × 1 mm., green
tinted reddish-brown at base. Corolla c. 30−40 mm. long, white, the tube usually
flushed deep pink above, the two upper lobes flushed pink outside, each with
3 magenta-pink stripes inside, the median stripe extending briefly into tube,
sometimes broken into dots, the three lower lobes each with a median magenta-
pink stripe, irregular magenta-pink spots on the palate and inside the tube, the

roof of the tube immaculate, bearded with long multicellular hairs and shorter gland-tipped ones, some minute unicellular clavate hairs on the floor of the tube; tube narrowly funnel-shaped, slightly curved, the floor conspicuously grooved, c. 20−25 mm. long, c. 3 mm. in diam. in lower half, widening above to 6−7 mm. across mouth; limb oblique, upper lip of 2 suborbicular lobes, 6 × 6 mm., lower lip 9−12 mm. long, lobes oblong, 6−8 × 5−6 mm. Stamens arising in middle third of corolla tube, filaments c. 6 mm. long, twisted, white, glandular above; anther lobes 1 mm. long, blue-black; staminodes conspicuous, c. 4 mm. long. Ovary c. 13−15 mm. long, densely pubescent with spreading, eglandular hairs; style c. 8 mm. long, similarly pubescent, dorsoventrally compressed, the apex bent over; stigma white, papillose, stomatomorphic. Capsule 25−50 × c. 2 mm., apiculate. Seeds 0.3−0.4 mm. long, reticulate.

Malawi. S: Mangochi Distr., Mangochi Mt., 1900−2000 m., fl. 21.xii.1963, *Chapman 2162* (MAL; SRGH). Mozambique. N: Lichinga, fl. & fr. 25.ii.1964, *Torre & Paiva 10803* (E; LISC; SRGH).
On mossy or wet rock faces at edge of evergreen forest; 1350−1800 m.

17. **Streptocarpus cyanandrus** B.L. Burtt in Notes Roy. Bot. Gard. Edinb. 24: 47 (1962).—
Hilliard & Burtt, Streptocarpus 245, fig. 37, N a & b, pl. 11 C (1971). Type: cult. in Hort. Bot. Reg. Edinb. ref. C3674 (E, holotype) from seed originally collected in Zimbabwe, Inyanga Downs, 2333 m., *Wild 4943*.

Perennial, all parts excluding corolla pilose with glandular and eglandular hairs. Leaves several, loosely associated in a tuft, young ones sessile, the flowering ones developing a conspicuous stalk (petiolode) up to c. 35 mm. long, lamina linear-oblong to lanceolate, 5−15 × 1−5 cm., green above, purple-red below, margin coarsely crenate-dentate, apex obtuse, base cuneate. Inflorescences 1 or 2 at junction of lamina and stalk, few-flowered. Peduncles up to 45 mm. long. Bracts inconspicuous. Pedicels up to 22 mm. long. Corolla c. 25−35 mm. long, glandular-pubescent outside, the roof of the tube bearded with glandular and eglandular hairs, the floor with minute unicellular hairs, magenta-pink above (colour variable?), paler below, each lobe with three magenta-pink stripes that extend down inside the tube in lines of dots, the floor of the tube profusely spotted magenta-pink; tube funnel-shaped, slightly curved, floor conspicuously grooved, c. 18−25 mm. long, 3−4 mm. in diam. in lower half, widening to 6−7 mm. across mouth; limb oblique, upper lip of 2 suborbicular to elliptic-oblong lobes 5−6 mm. long, lower lip 10−12 mm. long, lobes elliptic-oblong, 6−9 × 5−7 mm. Stamens arising in middle third of corolla tube, filaments 5−6 mm. long, glandular above; anther lobes 1 mm. long, blue-black; staminodes conspicuous. Ovary c. 6 mm. long, pubescent with glandular and eglandular hairs; style c. 10 mm. long with similar indumentum, dorsoventrally compressed, the tip bent over; stigma stomatomorphic, papillose. Capsule 15−18 mm. long, crowned with persistent style. Seeds 0.4−0.5 mm. long, reticulate.

Zimbabwe. E: Inyanga, World's View, fl. 20.ii.1965, *Plowes 2497* (E; SRGH).
Only known from the Inyanga area. Under rock overhangs; 2130−2400 m.

18. **Streptocarpus pumilus** B.L. Burtt in Kew Bull. 1936: 491 (1936).—Hilliard & Burtt, Streptocarpus 247, (1971). Type: Zimbabwe, Rusapi, 1.i.1935, *Eyles 8359* (K, holotype; SRGH).

Perennial. Leaves several, loosely associated in a tuft, young ones sessile, flowering ones becoming conspicuously stalked, linear-oblong to ovate, up to 80 × 40 mm., green above, flushed reddish-violet below, pilose on both surfaces, glandular hairs frequent below, margin crenate-dentate, apex obtuse, base subcordate to cuneate, stalk (petiolode) up to 40 mm. long, pilose. Inflorescences one or two at junction of lamina and stalk, few-flowered. Peduncles up to c. 40 mm. long; pedicels up to 15 mm. long, both wiry, sparsely pubescent. Bracts minute. Calyx divided to the base into 5 linear-lanceolate segments, 2 × 0.5 mm., sparsely pubescent with glandular and eglandular hairs. Corolla c. 12−15 mm. long, sparsely glandular-pubescent outside, the roof inside sparsely bearded with multicellular and unicellular hairs, white, tube flushed magenta-pink above,

magenta-pink median stripe on inside of each corolla lobe, floor and walls of
tube and base of lower lip profusely spotted magenta-pink; tube funnel-shaped,
slightly curved, 9 mm. long, c. 2—3 mm. across mouth; limb oblique, lower lip
c. 4—6 mm. long, all lobes elliptic-oblong, 3—4 × 1.5—2 mm. Stamens arising
in upper third of corolla tube; filaments c. 2 mm. long; anther lobes <1 mm.
long, blue-black at maturity; staminodes minute. Ovary c. 3 mm. long, pubescent; style c. 5 mm. long, dorsoventrally compressed; stigma stomatomorphic,
papillose. Capsule 7—10 mm. long, crowned with the persistent style base. Seeds
0.3—0.5 mm. long, reticulate.

Zimbabwe. E: Inyanga, Juliasdale, fl. & fr. 23.i.1973, *Biegel* 4163 (E; K; LISC; SRGH).
C: Domboshawa, 7.iii.1946, *Wild* 899 (K; SRGH).
In shelter of rocks on granite hills and on cave floors; c. 1500—1950 m.

19. **Streptocarpus hirticapsa** B.L. Burtt in Notes Roy. Bot. Gard. Edinb. 28: 211 (1968).—
Hilliard & Burtt, Streptocarpus 248, pl. 11, e (1971). Type: Zimbabwe, Chimanimani
Distr., Chimanimani Mts., "Stonehenge" Plateau, 1800 m., *Chase* 6908 (K, holotype;
SRGH).

Perennial, all parts excluding flower pilose with long spreading hairs. Leaves
several, loosely associated in a tuft, young ones sessile, flowering ones becoming
conspicuously stalked, stalk (petiolode) up to c. 35 mm. long, pilose, lamina
linear-oblong to almost ovate, variable both in size and shape, apparently not
exceeding c. 10 × 4 cm., base subcordate to cuneate, margin finely crenate-
dentate. Inflorescences 1 or 2 at junction of blade and stalk, few-flowered.
Peduncles up to c. 60 mm. long; pedicels up to c. 12 mm. long, both wiry. Bracts
minute. Calyx divided to the base into 5 linear-lanceolate segments c. 2 mm.
long, pilose. Corolla c. 13 mm. long, sparsely pubescent with glandular and
eglandular hairs, white, magenta-pink strip on each lobe and spotted magenta-
pink on floor of tube and base of lower lip; tube narrowly funnel-shaped, slightly
curved, 9 mm. long, c. 2 mm. across mouth; limb oblique, upper lip of 2 oblong-
elliptic lobes 3 × 2 mm., lower lip 5.5 mm. long, lobes oblong-elliptic, 4 × 2.5
mm. Stamens arising in upper third of corolla tube; filaments c. 2 mm. long;
anther lobes <1 mm. long; staminodes minute. Ovary 3 mm. long; style 6 mm.
long, dorsoventrally compressed; stigma stomatomorphic, papillose. Capsule
c. 8—15 × 1.5 mm. long, densely clothed in long spreading hairs, crowned with
the persistent style. Seeds 0.3—0.4 mm. long, reticulate.

Zimbabwe. E: Chimanimani, confluence Timbiri and Benzi rivers, fl. & fr. iv.1969,
Goldsmith 47/69 (E; K; LISC; NU; SRGH). **Mozambique**. MS: Chimanimani, both sides
Mozambique border, 7.vi.1949, *Wild* 2938 A (K).
In the shelter of overhanging rocks on mountain sides; 950—1800 m.

20. **Streptocarpus leptopus** Hilliard & Burtt in Notes Roy. Bot. Gard. Edinb. 28: 212 (1968);
Streptocarpus 252, fig. 37, P a & b (1971). Type: Malawi, Mulanje Mt., Great Ruo
Gorge, *Hilliard & Burtt* 4637 (E, holotype; NU).

Perennial, rosulate. Leaves several from a minute axis, prostrate, oblong-
lanceolate, shortly stalked, up to 150 × 20 mm., often much smaller, midrib
and lateral veins strongly raised below, depressed above, the leaf lamina ridged
between them, apex obtuse, base cuneate, margin crenate, dark green above,
often purplish below, pilose with spreading hairs on both surfaces. Inflorescences
one or two from the leaf stalk, generally 2-flowered, but up to 6. Peduncles very
slender and wiry, up to 75 mm. long, with a few spreading hairs. Bracts minute.
Pedicels up to 15 mm. long, wiry. Calyx divided to the base into 5 lanceolate
segments 3.5 × 1 mm., pubescent. Corolla 14 mm. long, white with relatively
large purple blotches on the palate and the floor of the tube, outside glabrous
or sparsely glandular-pubescent, inside with long unicellular clavate hairs on
floor of tube and palate; tube subcylindrical, slightly curved downwards and
upwards, 7mm. long, c . 3 mm. across mouth; limb oblique, upper lip of 2 erect
suborbicular lobes c. 1.5 mm. long, lower lip 6 mm. long, lobes oblong-elliptic,
c. 4 × 2.5 mm. Stamens arising in lower third of corolla tube, filaments 3.5
mm. long, twisted and slightly thickened upwards, anther lobes <1 mm. long;
staminodes minute. Ovary 1.5 mm. long, quite glabrous, with a few sessile

glands; style 4 mm. long, dorsoventrally compressed; stigma stomatomorphic, papillose. Capsule 7−10 × c. 2 mm., crowned with the persistent style. Seeds 0.4−0.5 mm. long, reticulate.

Malawi. S: Mt. Mulanje, foot of Great Ruo Gorge, 18.iii.1970, *Brummitt & Banda* 9194 (K; MAL; SRGH). **Mozambique.** Z: Malange, Serra Tumbine (Macinjiri), fl. & fr. 16.i.1971, *Hilliard & Burtt* 6272 (E; NU).
Only known from these two localities. Shady, moss-covered rocks; 870−1500 m.

21. **Streptocarpus rhodesianus** S. Moore in Journ. Bot. **49**: 188 (1911).− Burtt in Kew Bull. **1939**: 78 (1939).− Hilliard & Burtt, Streptocarpus 254 (1971). Type: Zambia, Katenina Hills, *Kassner* 2162 (K, holotype).
   Streptocarpus paucispiralis Engl., Bot. Jahrb. **15**: 217 (1921). Type: Zambia, Katenina Hills, *Kassner* 2162 (K, isotype).
   Streptocarpus rhodesianus var. *perlanatus* Duvign. in Bull. Soc. Roy. Bot. Belg. **96**: 180 (1963). Type from Zaire.

Perennial, rosulate. Leaves several, prostrate, elliptic-oblong or oblong, up to c. 20 × 19 cm., often much smaller, margin entire, base subcordate to cuneate, upper surface grey-green, lower often purplish, both surfaces almost lanate with long soft delicate hairs. Inflorescences several in succession from the base of the midrib, 2−10-flowered. Peduncles wiry, up to c. 10 cm. long; pedicels very slender, 7−15 mm. long, both glandular-pilose. Calyx divided to the base into 5 lanceolate segments c. 2 mm. long, glandular-pilose. Corolla c. 7−10 mm. long, sparingly gladular-pilose outside, tube "dull purple" or "dull wine red" outside, the lobes white, floor of tube white blotched wine red with 2 parallel bands of stout unicellular clavate hairs extending onto palate and lobes of lower lip, roof of tube uniform wine red, bearded with long delicate unicellular hairs; tube subcylindrical, widening slightly upwards, c. 8 mm. long, 3 mm. across mouth; limb oblique, lower lip c. 6 mm. long, the lobes oblong, c. 4 × 3 mm., 2 lobes of upper lip slightly smaller. Stamens arising in lower third of corolla tube, filaments c. 3 mm. long, twisted and thickened about the middle, purple, anther lobes c. 1 mm. long, white; staminodes minute. Ovary 2−2.5 mm. long, pilose; style 4 mm. long, slightly dorsoventrally compressed(?); stigma stomatomorphic, papillose. Capsule c. 5−10 × 1.5 mm., crowned with persistent style. Seeds 0.3 mm. long, reticulate.

Zambia. N: Kasama Distr., Malole Rocks, 1275 m., 1.iii.1960, *Richards* 12669 (E; K; NU; SRGH). W: Mwinilunga, 15 km. N. of Kalene Hill, 14.xii.1963, *Robinson* 6032 (K; SRGH). C: Mkushi Distr., Rira Gorge, 5.iv.1961, *Richards* 14949 (K; SRGH). E: Chipata Distr., Chanchenga Dam Protected Forest near Chadiza, 3.iii.1973, *Kornas* 3390 (K).
Also in Angola and Zaire. In shelter of rock outcrops on mountainsides; c. 1200−1500 m.

22. **Streptocarpus buchananii** C.B. Clarke in F.T.A. 4, 2: 510 (1906).−Burtt in Kew Bull. **1939**: 81 (1939). Hilliard & Burtt, Streptocarpus 331 (1971). Lectotype: Malawi, Shire Highlands, *Buchanan* 410 (E; K).
   Streptocarpus caulescens sensu C. B. Clarke in A. & C. DC. Monog. Phane. 5, 1: 154 (1883), quoad *Buchanan* 410 non Vatke.
   Streptocarpus lilacinus Engl., Bot. Jahrb. **57**: 214 (1921). Type from Tanzania.

Herb with succulent brittle stem, sparingly pubescent chiefly at the nodes, sometimes swollen at the base. Leaves opposite, more or less equal; petiole very variable in length, up to 50 mm. in the lower part of large plants, pubescent; lamina up to c. 90 × 45 mm., more or less elliptic (sometimes somewhat oblique), acute at the apex, more or less abruptly narrowed at the base, thinly pubescent on both surfaces, margins entire; lateral nerves c. 10 pairs, rather evenly spaced about 5 mm. apart, ascending. Inflorescences in the axils of the upper leaves; peduncles 10−15 cm., shortly pubescent; bracts c. 2 mm., linear, shortly pilose; pedicels 15−20 mm., clad with short eglandular and a few glandular hairs intermixed. Calyx segments narrowly triangular, 2 × 0.75 mm., pubescent with some glands near the base. Corolla c. 20 mm. long, deep blue; tube 10−12 mm., curved, sparsely glandular-pubescent outside; upper lip of two rounded lobes, 2−3 × 4−5 mm., probably erect; lower lip c. 10 mm. long, nearly 15 mm. across, median lobe c. 5 × 7 mm., laterals 4 × 7 mm., all broadly rounded. Stamens arising somewhat above the middle of the tube; filaments 3.5 mm., glabrous,

connective forming round pad at back of anther; anther lobes 0.75 mm.,
divergent. Disk annular, 0.5 mm. Ovary cylindrical, 5 mm. long, appressed-
pubescent and with some sessile glands underneath; style 4.5 mm., pubescent;
stigma capitate, papillose. Capsule slender, glabrescent (35) 50—70 mm. long.
Seeds 0.6 mm. long, verruculose.

Malawi. N: Chitipa Distr., Misuku Hills, Mughesse Forest, 1933 m., fl. & fr., 15.iv.1981,
Salubeni 3123 (E; MAL; SRGH). S: Zomba Mt., Mlunguzi Bridge, fl. 6.iii.1977, *Brummitt*
& *Seyani* 14801 (K; MAL).
   Also in S. Tanzania - Streamsides and damp forest margins.

124. BIGNONIACEAE

By M.A. Diniz

Trees and shrubs, sometimes straggling (sometimes climbers, very rarely herbaceous), mostly unarmed, not lactiferous. Stipules absent, rarely with pseudostipules (external scales of axillary shoots) well developed, sometimes foliaceous. Leaves usually opposite, rarely verticillate or alternate, imparipinnate or bipinnate (not in Flora Zambesiaca area), 1-jugate or simple, sometimes the terminal leaflet replaced by a tendril (not in Flora Zambesiaca area). Flowers zygomorphic, usually showy and in several-flowered terminal or axillary panicles or racemes, sometimes reduced to a fascicle or solitary flower. Calyx gamosepalous of various shapes, usually 5-dentate or -lobed or almost truncate or spathaceously split. Corolla gamopetalous bilabiate or not, with a conspicuous campanulate, infundibuliform or tubular tube and 5-lobed, limb usually imbricate in aestivation. Stamens epipetalous, usually 4, didynamous with 1 posticous staminode, not so often 5 equal stamens, very rarely 2 (not in Flora Zambesiaca area) adnate to the corolla tube, included or exserted; filaments slender, often dilated at the base; anthers 2-thecous, rarely 1-thecous (not in Flora Zambesiaca area), usually widely divergent or divaricate, dehiscing longitudinally. Disk hypogynous, nectariferous, annual or tubular, sometimes absent. Ovary superior, syncarpus 2-carpellary, typically bilocular with axile placentation or unilocular and with parietal placencation (*Kigelia*); ovules numerous in each locule, anatropous; style 1, terminal with bilamellate stigma. Fruit usually capsular, dehiscing by 2 loculicidal or septicidal valves, perpendicular or parallel to the septum, or fleshy and indehiscent (*Kigelia*). Seeds numerous, compressed and winged in the species with capsular fruits, wingless seeds in the species with fleshy fruits, without endosperm; embryo usually flat; cotyledons flattened, rarely folded, foliaceous.

About 120 genera and some 650 species distributed mostly throughout the tropical zones and a few genera in warm-temperate regions. From this predominantly woody family a few species are widely cultivated as ornamental, for their showy flowers. Various species of the genera are frequently planted along streets or in parks and gardens. Some of them have been occasionally found as escapes, on forest margins and roadsides.

Key to the cultivated species ★

1. Erect trees or shrubs (sometimes somewhat scandent in *Tecoma*) - - - 2
 − Climbers or lianas - - - - - - - - - - - 22
2. Leaves simple - - - - - - - - - - - - 3
 − Leaves digitate, trifoliate, once pinnate, bipinnate or tripinnate - - - 4
3. Leaves narrowly obovate to spathulate with attenuate base; corolla yellow with purple-brown markings; fruit spherical - up to 30 cm. diam. (*Calabash tree* - tropical America) - - - - - - - - - **Crescentia cujete** L.
 − Leaves ovate-deltoid with truncate to subcordate base; corolla creamy white with yellow and purple markings; fruit up to 40 cm. long, narrowly cylindrical, longitudinally ridged. (*Indian Bean Tree* - North America)
 - - - - - - - - - - **Catalpa bignonioides** Walt.
4. Leaves digitate or trifoliate - - - - - - - - - 5
 − Leaves once pinnate, bipinnate or tripinnate - - - - - - 10
5. Leaves digitate, branches without spines; corolla pale pink to magenta, with yellow in the throat, fading to white or cream, or corolla bright yellow with red lines into throat; fruit narrowly cylindrical 7−56 cm. long, attenuate. *Tabebuia* 4 sp - 7
 − Leaves trifoliate - - - - - - - - - - - 6
6. Branchlets with spines. Flowers solitary; corolla greenish white, sometimes with purple markings; fruit cylindrical up to 17 cm. long, curved, costate (Central America)
 - - - - - - - - - **Parmentiera aculeata** (Kunth) Seem.
 − Branchlets without spines. Flowers in a large branched inflorescence; corolla yellow with red lines into throat; fruit c. 7−32 cm. long (*Yellow* or *Golden Bells* - W. Indies, C. & S. America) - - - - - - **Tecoma stans** (L.) Juss. ex Kunth

★ By Sally Bidgood.

61

7. Leaflets pubescent at least on veins beneath, not lepidote; inflorescence tomentose, or villous with barbate or stellate hairs - - - - - - - - 8
 − Leaflets lepidote, without hairs; inflorescence lepidote; calyx lepidote, with or without a few scattered hairs - - - - - - - - - - - - - 9
8. Calyx tomentose; corolla crisped pubescent outside, pinkish purple to deep magenta, fruit glabrous, (S. America)
   - - - (Syn. *T. avellanedae*) **Tabebuia impetiginosa** (Mart. ex DC.) Standl.
 − Calyx villous with barbate or stellate hairs; corolla glabrous outside, yellow with red lines into the throat, fruit villous with branched or barbate hairs (S. America)
   - - - - - - - - - **Tebebuia chrysantha** (Jacq.) Nichols.
9. Leaflets narrowly obovate to elliptic with rounded to obtuse apex, frequently unequal at the base; corolla white to pinkish purple (*White Ceder* - West Indies)
   - - - - - - - - - - **Tabebuia pallida** (Lindl.) Miers
 − Leaflets elliptic with long-acuminate to cuspidate apex and obtuse to acute base; corolla white to pinkish purple or magenta (*Rosey Trumpet Tree* - C. and S. America)
   - - - - - - - - - - - **Tabebuia rosea** (Bertol.) DC.
10. Leaves once pinnate - - - - - - - - - - - - - 11
 − Leaves bipinnate or tripinnate - - - - - - - - - - 19
11. Calyx spathaceous - - - - - - - - - - - - - 12
 − Calyx campanulate or cylindrical - - - - - - - - - 15
12. Calyx pubescent - - - - - - - - - - - - - - 13
 − Calyx glabrous or glabrescent - - - - - - - - - - 14
13. Flowering calyx densely covered with multicellular hairs; corolla orange to red; fruit fusiform 15−29 cm., held erect (*African Tulip, Flame of the Forest* - Tropical Africa)
   - - - - - - - - - - - **Spathodea campanulata** Beauv.
 − Flowering calyx usually lepidote with a few scattered hairs, not densely covered with multicellular hairs; corolla yellow; fruit narrowly compressed 22−68 cm., frequently falcate, pendulous (Tropical Africa) (syn. *M. platycalyx, M. hildebrandtii*)
   - - - - - - - - - - **Markhamia lutea** (Benth.) K. Schum.
14. Corolla campanulate, 5-lobed, yellow, frequently with maroon or brown spots; calyx glabrescent, the surface usually lepidote; fruit narrow compressed 22−68 cm. long (Tropical Africa) - - - **Markhamia zanzibarica** (Bojer ex DC.) K. Schum.
 − Corolla unequally 4-lobed, the tube long, narrowly cylindrical, white or yellowish; fruit compressed, 7−32 cm. long (Mozambique)
   - - - - - - - - - **Dolichandrone alba** (Sim) Sprague
15. Calyx campanulate, large c. 1.4−2.1 × 1.7−2.2 cm., the lobes rounded to obtuse; leaflets entire (rarely crenate); corolla broadly campanulate, red, frequently with yellow in the throat; fruits narrow, up to 58 cm. long, compressed, loosely spiralled (Tropical Africa) - - - - - - - - - - **Fernandoa magnifica** Seem.
 − Calyx narrowly campanulate to cylindrical 0.3−1.7 × 0.2−0.8 cm.; leaflets crenate or serrate - - - - - - - - - - - - - - - 16
16. Leaflets crenate; corolla orange, rarely yellow, strongly bilabiate with anthers exserted, curved, c. 3.5−6(6.5) cm. of which the tube is c. 2−3.5 (4) cm.; fruit narrow compressed, attenuate at both ends 3.8−12.5 (19.5) cm. long (*Cape Honeysuckle* - Tropical Africa) *Tecomaria capensis* 2 subsp. - - - - - - 17
 − Leaflets serrate; corolla more or less regular, not strongly bilabiate; anthers more or less included straight c. 3.5−5.5 cm. narrowly campanulate with the tube constricted above the calyx for c. 0.3−0.5 cm., or infundibuliform; fruit narrow compressed c. 7−82 cm. long (West Indies, South and Central America) *Tecoma* 2 sp. - - - - - - - - - - - - - - - - 18
17. Calyx (0.4)0.5−0.8(1.0) cm., leaflets (2)3−4(5)
   - - - - - - **Tecomaria capensis** (Thunb.) Spach subsp. **capensis**
 − Calyx (0.8)1.0−1.8(2.3) cm., leaflets (3)4−5(6)
   - - **Tecomaria capensis** (Thunb.) Spach subsp. **nyassae** (Oliv.) Brummitt
18. Corolla narrowly campanulate constricted above the calyx for c. 0.3−0.5 cm. (*Yellow* or *Golden Bells* - S. U.S.A., Central and S. America)
   - - - - - - - - - **Tecoma stans** (L.) Juss. ex Kunth
 − Corolla infundibuliform (N. Argentina, Bolivia)
   - - - - - - - - - - **Tecoma tenuiflora** (DC.) Fabris
19. Leaflets numerous, 19 or more per secondary axis; corolla purple; fruits disk-like, (*Mimosa-leaved Ebony* - Brazil) - - - - **Jacaranda mimosifolia** D. Don
 − Leaflets up to nine per secondary axis; corolla white or very pale yellow-green; fruit narrowly compressed or cylindrical - - - - - - - - - - 20
20. Domatia present on the underside of the leaflets; calyx less than 0.4 cm. long; corolla white, tube very long and narrow, widening abruptly at the lobes; fruit narrow, compressed 22.5−35 × 1−2 cm. (*Indian Cork Tree* - Asia)
   - - - - - - - - - - **Millingtonia hortensis** L. f.
 − Domatia absent on the underside of the leaflets; calyx more than 0.7 mm. long; corolla yellow, white or very pale yellow-green within; fruit cylindrical, up to c. 68 cm. long (Asia) *Radermachera* 2 sp. - - - - - - - - - - - 21

21. Calyx 0.7—1.4 cm.; corolla c. 3.5—6.5 cm. campanulate; fruit verrucose
   -    -   -   -   -   **Radermachera xylocarpa** (Roxb.) K. Schum.
— Calyx 1.8—3.5 cm.; corolla (7.2) 8—13.5 (15) cm. infundibuliform; fruit slightly sulcate
   -    -   -   -   -   **Radermachera sinica** (Hance) Hemsl.
22. Leaflets (3) 5—15, tendrils absent . - - - - - - - - 23
— Leaflets 1—3, (rarely also with 2 stipule-like leaflets at the base of the petiole), tendrils present often, replacing terminal leaflet - - - - - - - - 26
23. Corolla (7) 8—11.5 cm., infundibuliform, pink, often cream on the lobes; inflorescences usually cauliflorus on old wood, sometimes on young foliage shoots; calyx (1.0) 1.5—2.5 cm., the lobes 0.4—0.6 cm.; leaflets (5.8) 6.5—13.5 × (2.1) 3—7.5 cm.; fruit not seen (S.E. Asia) - - - - - - **Tecomanthe dendrophila** (Bl.) K. Schum.
— Corolla 1—6 cm. - - - - - - - - - - - 24
24. Calyx 1—2 cm. with lobes up to 0.9 cm.; corolla 4—6 cm. broadly campanulate, pink, yellowish with dark pink guide lines in the throat; fruit 30—40 cm., cylindrical, compressed, apiculate (*Port St. John Creeper* - S. Africa)
   -    -   -   -   -   -   **Podranea brycei** (N.E. Br.) Sprague
— Calyx 0.2—0.9 cm. not obviously lobed, corolla 1—5 cm.; fruit c. 5.7—8 × 2—2.6 cm. ovoid ellipsoid acuminate *Pandorea* 2 sp. - - - - - - - 25
25. Calyx less than 0.4 cm., corolla 1—2.2 cm., creamy yellow with purple to brown markings on the inside (Australia, Polynesia)
   -    -   -   -   -   -   **Pandorea pandorana** (Andr.) Steenis
— Calyx 0.5—0.9 cm.; corolla 4—5 cm., white, pink or mauve, frequently with dark pink to crimson markings into throat (*Bower of Beauty* - Australia)
   -    -   -   -   -   - .**Pandorea jasminoides** (Lindl.) K. Schum.
26. Inflorescences terminal - - - - - - - - - - - , 27
— Inflorescences axillary - - - - - - - - - - - 33
27. Calyx glabrous or nearly so - - - - - - - - - - 28
— Calyx densely tomentose - - - - - - - - - - - 31
28. Calyx obviously toothed, the teeth subulate; flowers few per inflorescence, usually paired; corolla mauve to purple, lighter in the throat; fruit c. 7.5 cm. long × 3 cm. wide, shortly oblong, echinate (*Argentine Trumpet Vine* - S. America)
   -    -   -   -   -   **Clytostoma callystegioides** (Cham.) Bur. ex Griseb.
— Calyx not obviously toothed with subulate teeth, flowers several per inflorescence; fruit linear compressed - - - - - - - - - - - 29
29. Leaflets obovate, obtuse at the apex, cuneate at the base; calyx 0.7—1.2 cm. glabrous; corolla 5.5—9 cm., tubular campanulate, the lobes obovate, pink, mauve, magenta or purple, throat white or yellow, with or without purple lines; (S. America)
   -    -   -   -   **Saritaea magnifica** (Sprague ex Steenis) Dugand
— Leaflets ovate, rounded or acuminate at the apex, rounded, truncate or subcordate at the base - - - - - - - - - - - - 30
30. Corolla 2—4.5 cm., campanulate, pinkish mauve, glandular pubescent on the outside, lobes obovate, stamens included; calyx 0.7—1.2 cm.
   -    -   -   -   -   **Arrabidaea selloi** (Spreng.) Sandwith
— Corolla 5—8 cm., tubular, yellow, orange or red, densely tomentose on margins, lobes narrowly oblong to oblong, stamens exserted; calyx 0.4—0.7 cm. (*Flame Vine, Flaming Trumpet* - S. America) - - - - **Pyrostegia venusta** (Ker-Gawl.) Miers
31. Corolla 3—5 cm.; leaflets 2.5—4.5 × 2—4 cm.; corolla cream outside, yellow inside; fruit ellipsoid up to 24 (31) × 5—7 (7.5) cm., echinate (*Monkey's Hairbrush* - S. & Central America) (syn. *P. echinatum* (Jacq.) Baill.)
   -    -   -   -   **Pithecoctenium crucigerum** (L.) A. Gentry
— Corolla 6—11 cm.; lealets 5—14 × 2—7 cm. - - - - - - - 32
32. Calyx with large dark glands; corolla yellow; fruit 18—27.5 cm. long, up to 2.8 cm. wide (S. America) - - - - **Adenocalymma marginatum** (Cham.) DC.
— Calyx without large dark glands; corolla mauve to red or yellow with red markings and purplish lobes, yellow in the throat; fruit ellipsoid, obtuse at either end 14—18.5 × 4.8—7 cm. (Central America) - - **Distictis buccinatoria** (DC.) A. Gentry
33. Corolla pink to purple; leaflets coriaceous, the tertiary veins raised beneath; fruits c. 15.4 × 1.5—2 cm. longitudinally ridged, drying light brown (Brazil)
   -    -   -   -   **Mansoa difficilis** (Cham.) Bur. & K. Schum.
— Corolla yellow to orange red - - - - - - - - - - 34
34. Calyx large broadly campanulate 1—1.2 × 1.5—1.7 cm.; tendrils claw-like; leaflets small (2.7) 3.1—5.3 (5.6) × (0.8) 0.9—1.8 (2.3) cm.; corolla large 5.8—10 cm., × 5—7 cm. across the lobes, lobes very broadly obovate measuring about half the length of the tube; fruit narrow 39—153 cm. long 0.5—1.5 cm. wide, drying dark brown (*Cat's Claw* - Central and South America) - - **Macfadyena unguis-cati** (L.) A. Gentry
— Calyx <1 × 1 cm. - - - - - - - - - - - 35
35. Stamens exserted, corolla lobes narrowly oblong to oblong, densely tomentose on the margins; fruit up to 31.5 cm. long, narrowly compressed (*Flame Vine, Flaming Trumpet* - S. America) - - - - **Pyrostegia venusta** (Ker-Gawl.) Miers
— Stamens included, corolla lobes broadly oblong to very broadly oblong, galbrous or

with scattered hairs on the surface and margins · · · · · · 36
36. Flowers single or paired on short axillary dwarf shoots in the axils of the leaves; leaflets with sub-cordate to cordate base; fruit (only one specimen seen) 19.2 × 2 cm. long, compressed (*Cross Vine, Trumpet Flower* - North America)
· · · · · · · · · · · **Bignonia capreolata** L.
– Flowers in elongated axillary racemes (occasionally with leafy bracts at lower nodes; leaflets with cuneate base; fruit c. 9 × 6 cm., ellipsoid, compressed
· · · · · **Anemopaegma chamberlaynii** (Sims) Bur. & K. Schum.

Key to native genera

1. Flowers in pendulous lax panicles, up to 1 (1.5) m. long; fruit fleshy, sausage-shaped, up to 1 m. long, indehiscent; seeds not winged; ovary 1-locular · · **9. Kigelia**
– Flowers in erect and shorter panicles or racemes or fasciculate or solitary, fruit capsular, seeds winged; ovary 2-locular · · · · · · · · 2
2. Calyx spathaceous; leaves imparipinnate · · · · · · · 3
– Calyx lobed or toothed, not spathaceous; leaves imparipinnate, or 1-jugate or simple · · · · · · · · · · · · · · · 4
3. Corolla tube c. 5 cm. long, narrowly cylindrical, expanding in the throat; corolla white (in Africa only along the S. Mozambican coast) · · · **4. Dolichandrone**
– Corolla tube shorter, campanulate, corolla yellow or greenish-yellow mottled brownish-purple · · · · · · · · · · · · · · 5. **Markhamia**
4. Leaves simple, greyish woolly; corolla white, or whitish pink with a 4–8 cm. long cylindrical tube; capsule woody with warted valves · · 3. **Catophractes**
– Leaves imparipinnate, 1-jugate or simple; corolla if white (*Rhigozum*) with a less than 2 cm. long cylindrical tube; capsule neither woody nor warted · · · 5
5. Capsule cylindrical; corolla pale pink or orange to crimson with a yellow throat; leaves hysteranthous · · · · · · · · · · · · 6
– Capsule compressed; corolla of different colour; leaves and flowers coetaneous . 7
6. Corolla 3.5–5.5 cm. long, pinkish; flowers in terminal pubescent large panicles · · · · · · · · · · · · · 6. **Stereospermum**
– Corolla c. 8 cm. long, orange to crimson with a yellow throat; flowers in axillary glabrous short racemes · · · · · · · · · 8. **Fernandoa**
7. Leaves short, up to 5 cm. long, simple, 1-jugate or imparipinnate, clustered on cushions, perfect stamens 5; corolla deep yellow, pink or white, spiny shrubs or small · · · · · · · · · · · · · 2. **Rhigozum**
– Leaves large, 13–33 cm. long, imparipinnate; perfect stamens 4, corolla orange to scarlet or lavender; shrubs or small trees not spiny · · · · · 8
8. Corolla orange to scarlet with a curved narrowly cylindrical tube; stamens exserted; capsule usually 6–12 cm. long · · · · · · · · 1. **Tecomaria**
– Corolla lavander with a straight tube; stamens included; capsule usually more than 30 cm. long · · · · · · · · · 7. **Podranea**

1. TECOMARIA (Endl.) Spach

Tecomaria (Endl.) Spach, Hist. Nat. Veg. Phan. 9: 137 (1840).

Straggling shrubs or small tree. Leaves opposite, imparipinnate. Flowers in terminal racemes or racemose panicles. Calyx campanulate, 5-toothed. Corolla tube narrowly funnel-shaped, almost cylindrical, curved; corolla limb bilabiate, orange to scarlet. Stamens didynamous, exserted; thecae connate in the upper third, slightly divergent below. Disk cupular. Ovary bilocular; ovules numerous, 4-seriate in each locule. Capsule linear-oblong in outline, compressed, dehiscing at right angles to the septum. Seeds with membranous wings.

A monospecific African genus with two subspecies occurring in the Flora Zambesiaca area. Subsp. *capensis* is widely cultivated as ornamental in the tropics and subtropics.

Tecomaria capensis (Thunb.) Spach, Hist. Nat. Veg. Phan. 9: 137 (1840).—Seem. in Journ. Bot., Lond. 1: 21 (1863).—Spragus in F.T.A. 4, 2: 514 (1906).—Marloth, Fl. S. Afr. 3, 2: 151, t. 39 fig. A (1932).—Macnae & Kalk, Nat. Hist. Inhaca I., Mocamb.: 153 (1958).—F. White, F.F.N.R.: 380 (1962).—Paviani in Garcia de Orta 16: 166 (1968).—Palmer & Pitman, Trees of Southern Afr. 3: 2001 (1973).—Brummitt in Bull. Jard. Nat. Belg. 44: 421 (1974).—Compton, Fl. Swazil.: 537 (1976).—Libven in Fl. Afr. Centr., Bignoniaceae: (1977).—Palgrave, Trees of Southern Afr.: 827 (1981). Type from S. Africa.

Many-stemmed shrub with scandent branches, 1.5-4 m. tall, or small tree up to 6 (10) m. high. Bark pale brown, fissured; branchlets copiously lenticellate.

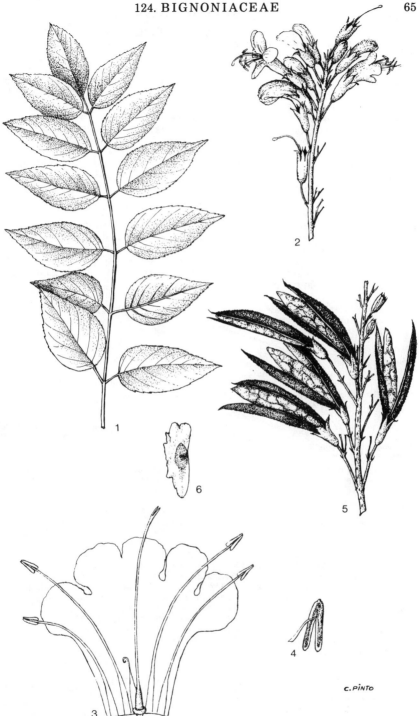

Tab. 10. TECOMARIA CAPENSIS Subsp. NYASSAE. 1, leaf (×½), *Andrada* 1818; 2, inflorescence (×½); 3, flower opened out showing view from inside and the gynoecium (×1); 4, stamen in dorsal view (×3), 2−4 from *Brummitt & Synge* 242; 5, branchlet with mature capsules (×½); 6, seed (×1), 5−6 from *Chapman* 1657.

Leaves 10−25(33) cm. long; petiole 2−7(9) cm. long, minutely pubescent to nearly glabrous, subwinged; rhachis narrowly winged, puberulous; leaflets 2−6-jugate subsessile or with the petiolules up to 1.2 cm. long, leaflet lamina 1.7−6.8(8.5) × 1.3−3(3.8) cm., elliptic, ovate or subcircular, glabrous to glabrescent on the superior surface, glabrous on the inferior, except towards the base on the midrib and usually densely pubescent on the axils of secondary nerves; apex acute, sometimes rounded, acuminate or mucronate, rarely retuse; base rounded or abruptly cuneate, asymmetric; margins crenate to toothed; lateral nerves 4−8 pairs. Inflorescence 10−32 cm. long. Peduncle up to 18 cm. long; bracts 3−7 mm. long, linear-subulate, rarely foliaceous, caducous; pedicels 4−11 (13) mm. long, minutely pubescent. Calyx 5−20 mm. long, lobed almost to the middle or more; lobes broadly triangular, minutely pubescent. Corolla tube c. 35 mm. long, narrowly cylindric widening near the mouth, curved; corolla bilabiate, the superior lip with the lobes connate about halfway, suberect, the inferior with the lobes free, recurved, 5 lobes 10−15 mm. long, ovate, ciliate. Stamens more or less equal in length, adnate to the corolla tube up to above the middle; filaments pubescent-glandular at the base; anthers 3−5 mm. long; thecae connate more or less to the middle, slightly divergent below. Disk 1.5 mm. in diam., lobed. Ovary c. 4 mm. long and 1 mm. across, oblong. Capsule up to 19.5 × 1.4 cm., flat, attenuate at both ends; valves smooth or rarely rugous. Seeds 6−10 × 13−33 mm. including the hyaline membranous wings, more or less rectangular in outline, numerous.

Leaflets (2)3−4(5)-jugate; calyx 5−8(10) mm. long, campanulate to shortly cylindric; calyx teeth less than 3 mm. long -   -   -   -   -   - subsp. *capensis*
Leaflets (3)4−5(6)-jugate; calyx (9)10−18 mm. long, cylindric; calyx teeth 3 mm. long or more -   -   -   -   -   -   -   -   - subsp. *nyassae*

Subsp. capensis
   Bignonia capensis Thunb., Prodr. Pl. Cap.: 105 (1800). Type as above.
   Tecomaria Petersii Klotzsch in Peters, Reise Mossamb., Bot.: 192 (1861). Type: Mozambique, Maputo, *Peters* s.n. (B, holotype †).
   Tecomaria krebsii Klotzsch, op. cit.: 8 (1861) in obs. Type from S. Africa.

Mozambique. GI: Between Inhambane and Inharrime, Nhacongo, fl. & fr. 22.vii.1954, *Barbosa 5549* (BM; LMA). M: Maputo, from Bela Vista to Tinonganine, fl. 6.viii.1957, *Barbosa & Lemos 777* (COI; K; LISC; SRGH).

Also in Swaziland and S. Africa (Transvaal Natal and Cape Prov.). In coastal dune scrubs, open woodlands and in thickets, also commonly cultivated as an ornamental hedge plant further north in the Flora Zambesiaca area; 0−600 m.

Subsp. **nyassae** (Oliver) Brummitt in Bull. Jard. Bot. Nat. Belg. **44**: 421 (1974).—Liben in Fl. Afr. Centr., Bignoniaceae: 12 t. 2 fig. C−D and t. fig. A−E (1977). TAB. 10. Type from Tanzania.
   Tecoma nyassae Oliver in Hook., Ic. Pl. 14 t. 1351 (1881). Type as above.
   Tecoma shirensis Baker in Kew Bull. **1894**: 30 (1894). Type: Malawi, Shire Highlands, *Buchanan 219* (BM, isolectotype; K, lectotype).
   Tecomaria nyassae (Oliver) K. Schum. in Engl. & Prantl, Nat. Pflanzenfam. 4 3b: 230 (1895).—Sprague in F.T.A. 4, 2: 515 (1906). Type as for *Tecoma nyassae*.
   Tecomaria shirensis (Baker) K. Schum. in Engl., Pflanzenw. Ost-Afr. C: 36 (1895).—Sprague loc. cit. Type as for *Tecoma shirensis*.
   Tecoma whytei C.H. Wright in Kew Bull. **1897**: 275 (1897). Type: Malawi, Zomba, Plateau, *Whyte* s.n. (K, holotype).
   Tecoma nyikensis Baker in Kew Bull. **1898**: 159 (1898). Type: Malawi, Nyika Plateau, *Whyte 112* (K, holotype).
   Tecomaria rupium Bullock in Kew Bull. **1931**: 274 (1931). Type from Tanzania.

Zambia. N: Mbala, fl. & fr.19.ix.1949, *Bullock 1001* (K; SRGH). **W**: Ndola fl. & fr. ix.1933, *Duff* (FHO). **C**: 1 km. NE. Lusaka fl. 13.vii.1930. *Huchinson & Gillett 3598* (BM; COI; K; LICS; SRGH). **E**: Lundazi Distr., Valley of Shire R., near summit of Kangampande Mt., Nyika Plateau 2100 m. fl. & fr. 3.v.1952, *White 2578* (K; FHO). **S**: near Mumbwa, Fl.1911 *Macaulay 779* (K). **Malawi. N**: Nchenachena Spur, Nyika Plateau, 2100 m., fl. 20.viii.1946, *Brass 17343* (BM; FHO; K; SRGH). **C**: between Dowa and Nkhota Kota 1500 m., fl. 22.vii.1936, *Burtt 6348* (BM; K). **S**: Mt. Mulanje, Nayowami Forest, 1950 m., fl. 23.viii.1956, *Newman & Whitmore 544* (BM; COI; SRGH). **Mozambique. N**: Lichinga, Serra Massangulo, 1450 m., fr. 5.iii.1964, *Torre & Paiva 11028* (C; LD; LISC; LMA; MO;

WAG). Z: Gúruè near tea plantation, fl. 19.ix. 1944, *Mendonça* 2119 (EA, n.v.; FHO; LISC; LMU, n.v.; LUAI, n.v.; SRGH). T: Angónia, Mt. Domuè, 1500 m., fl. 18.x.1943, *Torre* 6054 (BR, n.v.; LISC; WAG, n.v.).
Known also from Tanzania, Zaire (Haut-Katanga) and NE. Angola. In open woodland, evergreen forest, fringing *Widdringtonia* forest. *Brachystegia-Isoberlinia* woodland and on slopes of rocky hills; 900—2300 m.

T. capensis is widely cultivated in tropical areas and in warm temperate regions, for shaping edges or as ornamental shrub or tree.

2. RHIGOZUM Burch.

Rhigozum Burch. Trav. Int. S. Afr. 1: 299 (1822).

Spiny shrubs or small trees. Leaves solitary or clustered on cushions which are small lateral branchlets, simple 1-jugate or imparipinnate. Flowers borne in fascicles on lateral cushions or solitary, shortly pedicelate. Calyx short, campanulate, 5-toothed with or without hairs, opening early in bud. Corolla campanulate with 5 broad spreading lobes, bright yellow, white or pink. Stamens 5, slightly exserted, adnate up to the base of the enlarged part of the corolla tube; anther-thecae free only at the very base. Disk saucer-shaped, entire or lobed, thick. Ovary bilocular; ovules biseriate in each locule. Capsule oblong to elliptic compressed, splitting perpendiculary to the septum into 2 smooth valves. Seeds numerous, paper-winged.

A genus with 7 species, 5 of them S. African extending to tropical Africa, one from Madagascar and one from Ethiopia and the Horn of Africa (Afars and Issacs).

1. Corolla white or whitish-pink to pink; leaves simple; leaf margins undulate; branch in whorls of 3 - - - - - - - - - - - - 1. *trichotomum*
   − Corolla yellow; leaves simple or imparipinnate; leaf margins not undulate   - 2
2. Leaves usually simple (rarely some 1-jugate) clustered on shortly lanate cushions - - - - - - - - - - - - - 2. *brevispinosum*
   − Leaves imparipinnate or 1-jugate - - - - - - - - - 3
3. Leaves imparipinnate with leaflets (1)2—5-jugate; rhachis winged; petiole c. 4—9 mm. long - - - - - - - - - - - - - 4. *zambesiacum*
   − Leaves 1-jugate (rarely sometimes simple); rhachis not winged; petiole c. 2—6(8) mm. long - - - - - - - - - - - - 3. *obovatum*

1. **Rhigozum trichotomum** Burch., Trav. Int. S. Afr. 1. 299 (1822).—DC., Prodr. 9: 234 (1845).—Sprague in F.T.A. 4, 2: 530 (1906).-Bremek. & Oberm. in Ann. Transv. Mus. 16: 433 (1935).—Merxm. & Schreiber in Merxm. Prodr. Fl. SW. Afr. 128: 5 (1967).— Paviani in Garcia de Orta 16:167 (1968). Type from S. Africa (Cape Prov.).

Erect shrub 1—2(4) m. tall with ternate spiniform branches. Young branchlets greyish, at first with dense pubescence, with age glabrescent to glabrous. Leaves 7—20 × 2—4 mm., simple, solitary or clustered up to 10 on each cushion, spatulate to spathulate-oblong, rounded or slightly emarginate, rarely acute at the apex, tapering at the base, gland dotted, margins undulate, glabrous; sessile or shortly petiolate. Flowers usually terminal, rarely on lateral cushions, solitary or fascicled; pedicels 1—3 mm. long. Calyx 5—8 mm. long, irregularly 5-lobed; lobes more or less cuspidate, sometimes with prominent ribs, hairy to subglabrous, Corolla white, whitish-pink or pink; tube c. 17 mm. long, funnel-shaped, lower part cylindric and enclosed in the calyx, glabrous outside, hairy inside below the insertion of the stamens; lobes c. 10 mm., in diam., subcircular crenulate, spreading. Stamens adnate to the corolla tube, and up to 5—8 mm. from the base; filament 8—10 mm. long; anthers c. 4 mm. long, apiculate. Ovary 2—4 mm. long. Capsule 6—9 × 1—1.4 cm., long, beaked. Seeds not seen.

Botswana. SW: 14.5 km. S. of Ghanzi on Lobatsi Rd., 942 m., fl. 1.ii.1970, *Brown* 8299 (SRGH). SE: Khutse, 900 m., st. 24.iv.1972, *Standish-White* 389 (SRGH).
Also in Angola, Namibia and S.Africa (Orange Free State and Cape Prov.). In grasslands along riverbeds and pan fringes; 900—1330 m.

2. **Rhigozum brevispinosum** Kuntze in Jahrb. Konigl. Bot. Gart. Berl. 4: 270 (1886).—K. Schum. in Warb., Kunene-Samb.-Exped. Baum: 370 (1903), *"brevispinum"*.—Sprague in F.T.A. 4, 2: 531 (1906).—Merxm. & Schreiber in Merxm., Prodr. Fl. SW. Afr. 128:

5 (1967).—Paviani in Garcia de Orta, **16**: 168 (1968).—Palmer & Pitman, Trees of
Southern Afr. **3**: 2003 cum photogr. (1973).—Drummond in Kirkia **10**: 273 (1975).—
Palgrave, Trees of Southern Afr.: 828 (1981). Type from Namibia.
Rhigozum linifolium S. Moore in Journ. Bot., Lond. **37**: 172 (1899). Type from
Namibia.
Rhigozum spinosum Burch. ex Sprague in F.C. **4**, 2: 451 (1904). Type: Botswana,
Chue Spring, *Burchell* 2398/1 (K, holotype, LISC and MO, photo-holotype).

Erect rigid shrub or small tree, usually with patent spiny branches, 1.2−4
m. tall. Young branchlets quadrangular, grey-brownish, sometimes pubescent
or glabrescent. Leaves simple, rarely 1-jugate, sessile to subsessile, 20−50 ×
3−10 mm., oblanceolate to linear-oblanceolate, slender, rounded, retuse or
emarginate at the apex, tapering at the base, more or less pubescent when
young, glabrescent, margins entire, alternate on the young branchlets, 9−13
crowded together in each shortly lanate cushion on the older shoots; cushions
always below the short spines. Flowers single or clustered on the cushions;
pedicels 6−10 mm. long, pubescent. Calyx 5−9 mm. long, more or less 5-lobed,
pubescent outside, sometimes glandular in the upper part. Corolla golden-yellow,
sometimes with reddish streaks, with the tube 15 mm. long, minutely hairy
in the throat and below the insertion of the stamens; lobes 15−19 mm. wide,
subcircular, patent, crinkled, emarginate, margin ciliate. Stamens adnate up
to c. 10 mm. from the base corolla tube; filaments 7−12 mm. long; anthers 3−4.5
mm. long, not apiculate. Ovary 2−3 mm. long. Capsule 5.5−10 × 1.1−1.5 cm.
long, beaked; valves thin, light-brown. Seeds 1 × 1.8−2.2 cm., including the
wings.

Botswana. N: 20.75 km. N. of Shorobe Village, fl. 7.xii.1972 *Smith* 295 (K; SRGH). SW:
80 km. NE. Ghanzi along the Rd. to Maun, fl. 21.ix.1976, *Bergström* B−9 (K; SRGH).
SE: 3.22 km. NE. Derdepoort, c. 900 m., fr. 30.xi.1954, *Codd* 8899 (PRE; SRGH). **Zambia**.
S: Katima Mulilo, fr. 18.vi.1963, *Fanshawe* 7880 (FHO; K; LISC). **Zimbabwe**. W:
Nyamandhlovu Distr., fl. & fr. 14.xii.1950, *Orpen* 96/50 (K; LISC, SRGH). S: Gwanda
Distr., fl. 4.viii.1970, *Cleghorn* 2087 (K).
Also known from Angola, Namibia and S. Africa (Transvaal and N. Cape Prov.) In dry
open woodlands or open savanna-woodlands or on sandy soils, or in soils with limestone
outcrops; 700−1250 m.

3. Rhigozum obovatum Burch., Trav. Int. S. Afr. l: 389 (1822).—DC., Prodr. **9**: 234 (1845).—
Sprague in F.C. **4**, 2: 452 (1904).—Marloth, Fl. S. Afr. **3**: 151, t. 39-B and fig. 71
(1932).—Merxm. & Schreiber in Merxm., Prodr. Fl. SW. Afr. **128**: 5 (1967).—Palmer
& Pitman, Trees of Southern Afr. **3**: 2003 cum photogr. & 2 fig. (1973).—Palgrave
Trees of Southern Afr.: 829 (1981). Type from S. Africa (Cape Prov.).

Compact shrub or small tree 1−4.5 m. tall, with rigid spreading spiny and
greyish branches. Leaves 1-jugate rarely simple, fascicled on small cushions;
petioles 2−6(8) mm. long; leaflets 4−19 × 2−6 mm. obovate to oblong-obovate,
greyish-green, rounded often notched at the apex, tapering at the base, margins
entire revolute. Flowers usually in clusters of 1−3 on each cushion; pedicels
up to 5 mm. long. Calyx 3−4.5 mm. long, regularly 5-lobed; lobes short, rounded
minutely mucronate. Corolla bright yellow with a campanulate to funnel shaped
tube c. 15 mm. long, the lower cylindrical portion included in the calyx, outside
glabrous, inside pilose in the throat and below the insertion of the stamens;
lobes 9−16 mm. wide, subcircular, spreading, margin ciliate. Stamens adnate
up to c. 10 mm. from the base of the corolla tube; filaments 3−6 mm. long;
anthers 3.5−4 mm. long, slightly exserted. Ovary c. 1.5 mm. long; style 14−21
mm. long. Capsule 4.5−6.5 × 1−1.5 cm., white brownish, long, beaked. Seeds
1−1.5 × 1.5−2.5 cm. including the wings.

Zimbabwe. W: Matobo Distr., fl. 28.ix.1958, *Darbyshire* 2719 (SRGH). E: Mutare Distr.,
Sabi Valley, 3.2 km. S. of Rupisi Hot Springs, 600 m., fl.xi.1967, *Goldsmith* 114/67 (K;
LISC; SRGH). S: Gwanda Distr., Special Native Area "G", c. 550 m., fr. 15.xii.1956, *Davies*
2317 (K; SRGH).
Also in Namibia and S. Africa (Karroo Midlands, Orange Free State and N. and E. Cape
Prov.). In dry rocky places, *Mopane-Commiphora* scrub and basalt soils or granite
sandvelds; 150−1400 m.

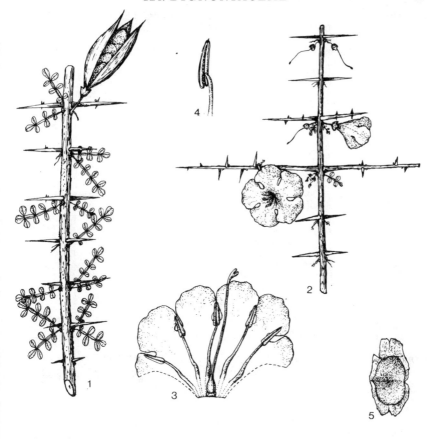

C.PINTO 83

Tab. 11. RHIGOZUM ZAMBESIACUM. 1, fruiting branchlet (×⅓), from *Torre & Correia* 18424; 2, flowering branchlet (×⅓); 3, corolla opened out showing stamens, viewed from inside and gynoecium (×1); 4, stamen (×2); 5, seed (×1), 2—5 from *Torre & Pereira* 12348.

4.Rhigozum zambesiacum Baker in Kew Bull. **1894**: 32 (1894).—K. Schum. in Engl. Pflanzenw. Ost-Afr. C: 363 (1895).—Sprague in F.T.A. 4, 2: 532 (1906).—Paviani in Garcia de Orta 16: 168 (1968).—Palmer & Pitman, Trees of Southern Afr. 3: 2005 cum fig. (1973).—Drummond in Kirkia, 10: 273 (1975).—Compton, Fl. Swazil.: 538 (1976).—Palgrave, Trees of Southern Afr.: 829 (1981). TAB. 11. Type: Mozambique, near Tete, fl. & fr. xi.1858, *Kirk* s.n. (K, holotype; LISC; MO, photo-holotype).

Spiny shrub or small tree up to 4 m. tall (reaching 7 m. outside Flora Zambesiaca area), with rigid spreading branches with smooth greyish-brown bark, with shallow longitudinal ridges. Leaves 13—40 mm. long, (2) 3—5-jugate, very rarely 1-jugate, solitary or clustered in small almost glabrous cushions situated below the spines; petiole 4—9 mm. long; rhachis winged; leaflets 5—9 (14) × 3—4.5(6) mm., ovate, obovate, elliptic or rounded, sometimes emarginate at the apex, narrowly tapering or broadly acute at the base, margins entire, sessile. Flowers solitary or clustered on the cushions; pedicels 2—5 mm. long. Calyx 3—5 mm. long, subglabrous, 5-lobed, usually pilose at the apex. Corolla golden-yellow, with the tube 10—17 mm. long, glabrous lower cylindrical part slightly exceeding the calyx; lobes 8—13 mm., subcircular, crenulate, margin ciliate, puberulous on the inner surface at the sinuses, becoming glabrous at last. Stamens adnate up to c. 8 mm. from the base of the corolla tube; filaments c.5 mm. long; anthers 4—5.5 mm. long, slightly apiculate, exerted from the corolla tube. Ovary c. 2mm. long; style 11−17 mm. long. Capsule 3.5−8 × 1.2−2 cm. beaked, pale brown. Seeds 1−1.3 × 1.7−2.2 cm., including the wings.

Zimbabwe. E: Chipinge Distr., Mutandawhe, fl. 12.x.1972, *Cannell* 529 (SRGH). S: Chiredzi Distr., Triangle Sugar Estate, fr. 22.i.1949, *Wild* 2788 (K; LISC; SRGH). **Mozambique.** MS: Sofala Prov., Chemba, 24 km. from Chambara to plateau of serra Lupata Mts., Nhamacolongo, fl. 14.v.1971, *Torre & Correia* 18424 (BR; COI; LISC; LMA; LUA; M). GI: Guijá, fr. 27.vi. 1947, *Pedro & Pedrogão* 2115 (LMA; LMU). M: Magude, between Maéle and Uanitze, fl. & fr. 30.x.l944, *Mendonça* 3157 (COI; LISC; SRGH).

Also in S. Africa (N. Transvaal and N. Zululand) and Swaziland. In deciduous woodlands, coastal thickets, dry sandy ground and *Colophospermum mopane* scrub; 0−500 m.

3. CATOPHRACTES

Catophractes D. Don in Proc. Linn. Soc. Lond. 1: 4 (1839).

Erect spiny shrub or small tree with dense grey woolly hairs. Leaves opposite, often fascicled, simple. Flowers solitary or fascicled, axillary more or less dense in the uppermost part of the branches. Calyx tubiform, shortly splitting at one side with 5(6) filiform lobes. Corolla campanulate, slightly bilabiate with 5(6−7) patent lobes; tube cylindrical, rather long Stamens as many as the corolla lobes, adnate to the corolla tube, perfectly exerted; anther-thecae connate in the upper part, free and parallel below. Ovary bilocular, ovoid surrounded by a cupular disk; ovules few, biseriate in each locule. Capsule elliptic, thick woody walled, rather flattened parallel to the septum, dehiscing along the flat faces; valves warty, tomentose-greyish and at the end dark brown. Seeds many, papery winged.

A monospecific genus from sub-tropical and Southern Africa.

Catophractes alexandri D. Don in Proc. Linn. Soc. Lond. 1: 4 (1839); in Trans. Linn. Soc. Lond., Bot. 18: 307, t. 22 (1841).—DC., Prodr. 9: 233 (1845).—K. Schum. in Engl. & Prantl, Nat. Pflanzenfam. 4, 3b: 233 (1894).—Sprague in F.T.A. 4, 2: 533 (1906).— Bremek. & Oberm. in Ann. Transv. Mus. 16: 433 (1935).—Merxm. & Schreiber in Merxm., Prodr. Fl. SW. Afr. 128: 2 (1967).—Paviani in Garcia de Orta 16: 169 (1968).— Drummond in Kirkia 10: (1975).—Palmer & Pitman, Trees of Southern Afr. 3: 2006−7 cum 5 photogr. (1973).—Palgrave, Trees of Southern Afr.: 830 (1981). TAB.12. Type from Namibia.
    Catophractes welwitschii Seem. in Journ. Bot., Lond. 3: 331, t. 39 (1865) *"Welwitschii"*. Type from Angola.

Spiny shrub or small tree up to 4 m. high, much-branched from the base with divaricate woolly branches when young, becoming glabrous with age. Leaves petiolate; lamina 0.9−3.5 (4.7) × 0.5−1.7(2.3) cm. oblong to ovate or obovate, chartaceous to subcoriaceous, woolly on both surfaces greyish-green on the superior surface, more whitish on the inferior, apex rounded or rarely acute, base narrowly rounded to broadly cuneate, midrib and lateral nerves slightly sunken above, very prominent below as well as all the net-veining; margins irregularly toothed to crenate with small subcircular submarginal glands mainly from the base; petiole 1−5(7) mm. long. Flowers white, showy, in few-flowered fascicles or solitary on the young branches and in leaf axils; pedicels up to 5 mm. long, thick woolly. Calyx 18−40(47) mm. long, 5−9 mm. in diam. shortly splitting at one side conspicuously 5−6 ribbed; lobes 1−5(10) mm. long, filiform, woolly on the upper face, sparsely pilose in the lower. Corolla tube 4−8.5 cm., long, narrow, greenish, glabrous outside, hirsute inside below the insertion of the stamens, or not; corolla lobes 1.5−3 cm. in diam., subcircular. Filaments adnate to the corolla tube for all their length or free on the upper part; anthers 4−6 mm. long. Ovary c. 2.5 mm. long, scabrous with thick dilated-base hairs; style 5.5−7.5 cm. long, glabrous or sparsely hairy on the lower half, thickened to a bilobed stigma. Capsule 3−6 × 1−2.4 cm., shortly or longly beaked, stipitate; valves warty, densely greyish-green tomentose. Seeds c. 1.5 × 2.5 cm. brownish including the hyaline membranous wing.

Botswana. N: Ngamiland, near Nata R., Maun Rd.,fl. & fr. 9.iii.1961, *Richards* 14630 (K; SRGH). SW: Masettheng Pan, E. side, fl. & fr. 16.xii.1976, *Bergstrom* 28 (K; SRGH).

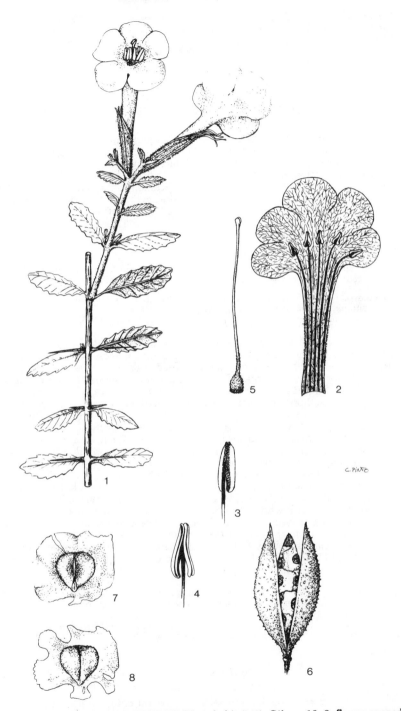

Tab. 12. CATOPHRACTES ALEXANDRI. 1, habit (×⅓), *Gibson* 13; 2, flower opened out showing stamens (×⅓); 3, dorsal side of stamen (×2); 4, ventral side of stamens (×2); 5, gynoecium (×⅓); 6, capsule opened (×⅓); 7, seed in basal (or ventral) view (×1); 8, seed in basal view (×1), 2−8 from *Torre* 8459.

SE: 9 km. NE. Khutsa, fl. & fr. 25.v.1972, *Coleman* 90 (K; LISC: SRGH). **Zimbabwe**. S: Gwanda Distr., 37 km. NW. Beitbridge, fl. 20.xii.1962, *Leach* 11550 (K; PRE).
Also in Angola, Namibia and S. Africa (Transvaal). In *Colophospermum mopane* and *Acacia* woodlands, usually on sandy soils or in xerophyte scrubs on limestone ridges and outcrops; up to 1200 m.

4. DOLICHANDRONE (Fenzl) Seem.

Dolichandrone (Fenzl) Seem. in Ann. Mag. Nat. Hist., Ser. 3, **10**: 31 (1862), nom. conserv.

Unarmed shrubs or trees. Leaves opposite, imparipinnate, deciduous. Flowers born in terminal panicles, racemes or racemose panicles, fragrant, opening during the night. Calyx spathaceously splitting along the posterior side, beaked at the apex. Corolla campanulate; tube narrowly cylindrical long exceeding the calyx; lobes 5, subequal, wavy at the margin. Stamens 4, didynamous, anthers included, introrse, glabrous; staminode present, very small. Disk annular, thick. Ovary oblong, cylindric, bilocular; ovules many, pluriseriate. Capsule narrowly linear in outline, terete or compressed parallel to the false septum, straight or curved bivalved loculicidally dehiscent; seeds flat, winged, 1-several on each side of the wing.

A genus comprising some 8 species distributed in tropical Asia and Australia and one species occurring in tropical Africa, known only from Mozambique.

Dolichandrone alba (Sim) Sprague in Kew Bull. **1919**: 308 (1919). TAB. 13. Type: Mozambique, *Sim* 5244 (destroyed? ★).
Spathodea alba Sim, For. Fl. Port. E. Afr. 92, t. 75B (1909). Type as above.

Deciduous shrub or small tree 3−12 m. tall. Bark pale brown to greyish longitudinally finely rough. Branchlets puberulous, soon becoming glabrous except at the leaflet articulations and provided with small scales which persist on the inferior surface of the adult leaves. Leaves opposite, imparipinnate, (1)3−4(5)-jugate; leaflets 1.5−7.5 (8.7) × 1−4(5) cm. ovate, elliptic or obovate, the more proximal ones smaller, sessile or subsessile, except the terminal one which is petiolulate, apex rounded to acuminate, base broadly cuneate to rounded, margin entire to minutely serrate; rhachis (2.2)2.7−14.5 cm. slightly sulcate above, terete below; petiole 0.5−3.7(4.7) cm. long, pseudostipules 0.3−1 cm. long, ovate. Inflorescence up to 48 cm. long, many flowered, but only 2−3 flowers opening at the same time; pedicels 1−1.7 cm. long. Calyx 2−3.3 cm. long coriaceous, spathaceous, slightly recurve-beaked, splitting along the posterior side. Corolla tube (3.5) 4−5(5.5) cm. long, narrow, slightly widening at the upper part, glabrous outside, more or less brownish pilose inside towards the base; corolla lobes white, 2−3 × 1.7−2.5 cm. rounded, somewhat fleshy, wavy at the margin. Stamens didynamous; subsessile, included, adnate up to about four fifths from the base of the corolla tube; anthers c. 3mm. long. Disk 2 mm. long, annular. Ovary c. 5 mm. long, longitudinally bisulcate. Capsule 17−38 (50) cm. long, 1.3−2.8 cm. across and c. 0.5 cm. thick. Seeds 0.6−1 × 2.5−4.2 cm., including the wing.

Mozambique. GI: Gaza, 30 km. of Manjacaze, fl. & fr. 26.i.1941, *Torre* 2560 (FHO; K; LISC; SRGH). M: Maputo, Av. Afonso de Albuquerque, near the Swiss Mission, fl. 4.i.1941, *Hornby* 4105 (LMA).
Known only from S. Mozambique. In dry deciduous woodland, fringing forests or thickets, on sandy soils, mainly near the coast; low altitude.

5. MARKHAMIA Seem. ex Baill

Markhamia Seem. ex Baill. in Hist. Pl. **10**: 47 (1888) ″1891″.

Shrubs or trees. Leaves opposite, imparipinnate; pseudostipules well developed. Inflorescence a terminal or axilliary panicle or raceme. Calyx

★ In the sequence of inquiries made to BOL, GRA, J, NH, NU, PRE, SAM and STE we got from NH the information that: ″most of *Sims* collections were destroyed by his wife after he died″.

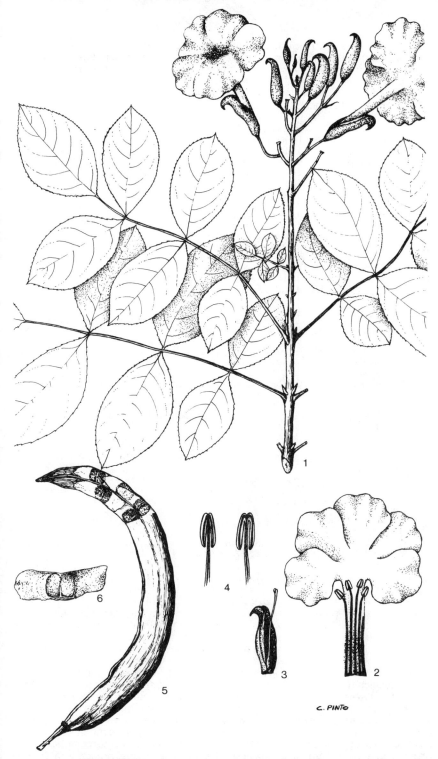

C. PINTO

Tab. 13. DOLICHANDRONE ALBA. 1, habit (×⅓); 2, corolla opened out showing stamens and staminode viewed from inside (×⅓); 3, calyx opened, showing gynoecium (×⅓); 4, stamens, two views (×3), 1—4 from *Torre* 6768; 5, capsule with part of one valve removed, showing seeds (×⅓); 6, seed (×1), 5—6 from *Gomes e Sousa* 3984.

spathaceous, cuspidate or uncinate at the apex, shorter than the corolla tube. Corolla campanulate, 5-lobed, bilabiate, funnel-shaped at the base. Stamens didynamous, enclosed; anther-thecae connate above, divergent towards the base. Disk annular or cup-shaped. Ovary oblong, bilocular; ovules numerous 4—6-seriate per locule. Capsule long, flattened parallel to the false-septum, bivalved, valves splitting loculicidally. seeds paper-winged.

Pseudostipules folaceous, subcircular to reniform; calyx 1—1.5 cm. long; corolla yellow-greenish flecked with maroon; corolla tube 2—3 cm. long; capsule hairless, lenticellate, sparsely lepidote - - - - - - - - - - 1. *zanzibarica*
Pseudostipules ovate to subulate-acuminate; calyx 1.9—3.3 cm. long; corolla bright yellow with brown-reddish striate on the lower lobes; corolla tube 2.5—4.5 cm. long; capsule minutely tomentose to velvety, neither lenticellate nor lepidote
- - - - - - - - - - - - - - 2. *obtusifolia*

1. **Markhamia zanzibarica** (Bojer ex DC.) K. Schum. in Engl., Abh. Königl. Preuss. Akad. Wiss. Berl. **1894**: 16 (1894); in Engl. & Prantl, Nat. Pflanzenfam. 4, 3b: 242 (1895); in Engl., Pflanzenw. Ost-Afr. C: 363 (1895).—Sprague in F.T.A. 4, 2: 523 (1906).— Dale & Greenway, Kenya Trees & Shrubs: 64 (1961).—Paviani in Garcia de Orta 16: 171 (1968).—Liben in Fl. Afr. Centr., Bignoniacea: 26 (1977). TAB. 14 fig. A. Type from Tanzania (Zanzibar).
    Spathodea zanzibarica Bojer ex DC., Prodr. 9: 208 (1845).—Klotzsch in Peters, Reise Mossamb., Bot. I: 191 (1861). Type as above.
    Spathodea acuminata Klotzsch in Peters, loc.cit. Type: Mozambique, Tete, Rios de Sena *Peters* s.n. (B†).
    Spathodea puberula Klotzsch in Peters, tom. cit.: 192 (1861). Type as above.
    Muenteria stenocarpa Welw. ex Seem. in Journ. Bot. 3: 329, t. 36 (1865). Syntypes from Angola.
    Spathodea stenocarpa Welw. in Seem. loc. cit., pro synom. Syntypes as above.
    Muenteria puberula (Klotzsch) Seem. in Journ. Bot. 8: 339 (1870). Type as for *Spathodea puberula*.
    Dolichandrone hirsuta Baker in Kew Bull. **1894**: 31 (1894). Type: Mozambique, Tete, banks of lower Zambesi, ix.1858, *Kirk* s.n. (K, holotype).
    Dolichandrone latifolia Baker loc. cit. Type from Kenya.
    Dolichandrone stenocarpa (Seem.) Baker loc. cit. Syntypes as for *Spathodea stenocarpa*.
    Markhamia acuminata (Klotzsch) K. Schum. in Engl., Pflanzenw. Ost-Afr. C: 363 (1895).—Sprague in F.T.A. 4, 2: 524 (1906).—N.E. Br. in Kew Bull. **1909**: 126 (1909).— Bremek. & Oberm. in Ann. Transv. Mus. 16: 433 (1935).—Brenan, T.T.C.L. 5, 2: 71 (1949).—Pardy in Rhod. Agric. Journ. 53: 58 cum 2 photogr. (1956).—Williamson, Useful Pl. Nyasal.: 81 (1955).—F White, F.F.N.R.: 379 (1962).—Merxm. & Schreiberin, Merxm., Prodr. Fl. SW. Afr. 128: 3 (1967).—Palmer & Pitman, Trees of Southern Afr. 3: 2008—9 cum 3 photogr. & fig. (1967).—Paviani in Garcia de orta 16: 171 (1968).—Palgrave, Trees Southern Afr.: 831 (1981). Type as for *Spathodea acuminata*.
    Markhamia puberula (Klotzsch) K. Schum. in Engl. & Prantl, Pflanzenfam. 4, 3b: 242 (1895).—Sprague in F.T.A. 4, 2: 523 (1906). Type as for *Spathodea puberula*.
    Markhamia stenocarpa (Seem.) K. Schum. in Engl. & Prantl, Nat. Pflanzenfam. 4, 3b: 242 (1895).—Sprague in F.T.A. 4, 2: 524 (1906). Syntypes from Angola.

Shrub 2—5 m. tall or a small often straggling tree up to 9 m. tall. Bark grey, smooth or rough, peeling off soon. Young branchlets minutely lepidote, sometimes with conspicuous lenticels. Leaves up to 35 cm. long (1) 2—4-jugate, size of the leaflet-pairs increasing progressively from the base; pseudostipules 0.5—1.7(2) cm. in diam., subcircular to reniform; petiole 2—7(9) cm. long, flat above, sometimes slightly winged, terete below, leaflet lamina 2—24.5(32.5) × 2—13 cm., elliptic, obovate or almost subcircular, sessile or with petiolules up to 5 mm. long, acute, acuminate to longly acuminate, rarely obtuse at the apex, tapering towards the often asymmetric base, pubescent and minutely scaly at both surfaces, with age becoming minutely and sparsely puberulous or even glabrous; lower surface with pubescent axillary domatiae more or less conspicuous and sometimes with small circular black glands near and on both sides of the midrib; lateral nerves 6—12(14), impressed above and prominent below; margins entire or finely toothed. Inflorescence a terminal or axillary panicle or raceme rather lax, 5—20(23) cm. long, scaly glabrous or puberulous; pedicels up to 1.5(2) cm. long, 2-bracteate below the middle; bracts 2—5(7) mm. long, triangular-acuminate, ciliate at the margins. Calyx 10—15(19) mm. long cuspidate or uncinate splitting at one side down to 8 mm. from the base

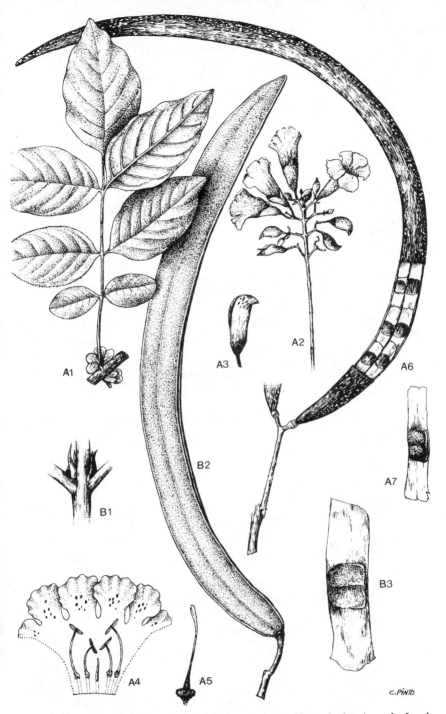

Tab. 14. A.—MARKHAMIA ZANZIBARICA. A1, part of branch showing a leaf and
foliaceous pseudostipules (×½), from *Barbosa* 2569; A2, inflorescence (×½); A3,
spatheceous calyx (×1); A4, corolla opened out showing stamens and staminode,
viewed from inside (×1); A5, gynoecium (×1), A2−5 from *Levy* 74; A6, capsule with
seeds (×½); A7, seed (×1), A6−7 from *Davidson* 54. B.—MARKHAMIA
OBTUSIFOLIA. B1, part of branch showing the pseudostipules (×1); B2, capsule
(×½); B3, seed (×½), all from *Macedo* 1302.

sometimes provided with scattered glands towards the apex and opposite to
the fissure. Corolla funnel-shaped to campanulate, tube (18) 20—30(43) mm. long,
yellow-greenish flecked with maroon; lobes 10—15 mm. in diam., subcircular,
sometimes with conspicuous small glands near the mouth. Stamen-filaments
9—14 mm. long adnate to the corolla tube up to c. 5 mm. from the base, corolla
tube provided with pluricellular hairs at the insertion points of the filaments;
anther-thecae c. 1.5 mm. long, divergent. Disk 1.5 mm. long and 2—3 mm. in
diam. Ovary 3—5 mm. long, sometimes lepidote; style 15—27 mm. long. Capsule
slender, 22—68 × 0.9—1.5 cm., straight or slightly falcate, glabrous, lenticellate.
Seeds 4—6 × 20—40 mm. including the wing.

Botswana. N: Ngamiland, Mwakupan area, fl. & fr. immat. 7.iii.1969. *De Hoogh* 137
(K; SRGH). SW: c. 2.5 km. W. of Kubi Gate, fl. & immat. fr. 22.ii.1970, *Brown* 8689
(SRGH). SE: Tuli Block, Merryhill Farm, fl. & immat. fr. 20.xii.1971, *Stephen & Wilson*
498 (PRE). **Zambia.** B: Sesheke Distr., Simongora Forest Reserve, 16 km. S. of Masese,
fr. 26.i.1952, *White* 1972 (FHO; K). W: Kitwe, forest nursery, st. s.d., *Fanshawe* 7688
(K). C: Mt. Makulu, near Chilanga, fl. l.i.1957, *Angus* 1810 (FHO; K; SRGH). S: Choma
Distr. Mapanza, 1050 m, fl. 2.xii.1958, *Robinson* 2939 (K; PRE). **Zimbabwe.** N: Urungwe
Distr., Msukwe R., 900 m., fl. 17.xi.1953, *Wild* 4162 (K; LISC). W: 2.5 km. W. of Binga,
460 m., fl. 8.xi.1958, *Phipps* 1410 (K; SRGH). C: Sebakwe, 1200 m., fl. xii.1904, *Eyles*
171 (BM; SRGH). E: Chimanimani Distr., Rd. to Birchenough Bridge near junction to
Biriwiri, fl. & fr. 15.xii.1963, *Chase* 8086 (SRGH). S: N. bank of Lundi R., Chipinda Pools
area, fl. 2.xii.1959, *Goodier* 691 (K; LISC; SRGH). **Malawi.** S: Sokola N.A. Nankumba,
Mangochi, fl. 19.xi.1954, *Jackson* 1387 (FHO; K). **Mozambique.** N: Imala, near Muíte,
Rd. Macubúri, fl. & fr. 21.xi.1936, *Torre* 1050 (COI; LISC). T: Estima-Tete, near Cahó
village, M'senangoé, fr. 27.i.1972 *Maçedo* 4710 (LISC; LMA; SRGH). MS: Sofala,
Gorongosa, between Zangorga and Kanga-Wihole Rd. Vila Paiva de Andrada, fl. & fr.
19.xi.1956, *Gomes e Sousa* 4329 (COI; FHO; K; LMA; SRGH). GI: Gaza, Limpopo, fl.
xii.1928, *Hutchinson* 2117 (K).
   From Angola, Zaire and Kenya to S. Africa. In dense forests, *Colophospermum mopane*
and *Brachystegia* woodlands, mixed forests, *Acacia* savanna woodlands, and riverine
formations; 0-1500 m.

2. **Markhamia obtusifolia** (Baker) Sprague in Kew Bull. 1919: 312 (1919).—Williamson,
      Useful Pl. Nyasal.: 81 (1955).—Dale & Greenway, Kenya Trees & Shrubs: 62 (1961).—
      F. White, F.F.N.R.: 379 (1962).—Gomes e Sousa, Dendrol. Moçamb. 2: 663 (1967).—
      Merxm. & Schreiber in Merxm., Prodr. Fl.SW. Afr. 128: 3 (1967).—Paviani in Garcia
      de Orta 16: 172 (1968).—Liben in Fl. Afr. Centre., Bignoniaceae: 32, t. 8. A-C (1977).—
      Palgrave, Trees of Southern Afr.: 831 (1981). TAB. 14 fig. B. Syntypes from Tazania,
      Malawi and Mozambique: Chupanga, *Kirk* s.n. (K, lectotype and isolectotype here
      designated; LISC, photo).
      Dolichandrone obtusifolia Baker in Kew Bull. 1894: 31 (1894).—Lectotype as above.
      Markhamia lanata K. Schum. in Engl. & Prantl, Nat, Pflanzenfam. 4, 3b: 242
      (1895).—Sprague in F.T.A. 4, 2: 527 (1906). Type uncertain (? B†).
      Markhamia paucifoliolata De Wild. in Ann. Mus. Congo Belge Bot., Ser. 4, 1: 131
      (1903). Type from Zaire.
      Markhamia verdickii De Wild., tom. cit.: 132 (1903). Type from Zaire.

Bushy shrub up to 1.5—5 m. or small tree 5—15 m. high. Bark light brown
to grey, smooth to somewhat striated in large specimens. Branchlets velvety
tomentose with golden brown hairs, soon glabrescent. Leaves 18—56 cm. long
(1)3—5(6)-jugate, inferior pairs smaller, pseudostipules 0.5—1.3 cm. long,
subulate-acuminate, tomentose; petiole 2—9.5 cm. long; leaflet-lamina
(4.5)6.5—5.17(23) × 2.5—9.5(13.5) cm.; elliptic to oblong, ovate or obovate,
sessile, apex obtuse to rounded, sometimes acute to shortly acuminate, rarely
longly acuminate, base obtuse, rounded to subcordate, sometimes asymmetric
margins entire or rarely thinly serrate pubescent to puberulous on the superior
surface, densely woolly on the inferior; lateral nerves (6)7—12(14). Inflorescence
a dense and tomentose terminal, many-flowered densely branched panicle
10—30(40) cm. long; pedicels 1—2.3(2.8) cm. long; bracts 0.5—1(1.5) cm. long.
Calyx 1.9—3.3 × 1.5—2.5 cm. obtusely cuspidate or uncinate, fissuring down
to 4—8 mm. from the base, densely tomentose with golden hairs. Corolla bright
yellow, funnel-shaped to campanulate; lobes spreading, subcircular, 1.7—3 cm.
in diam., the 3 lower ones streaked brown-reddish and with conspicuous glands
near the mouth of the 2.5—4.5 cm. long corolla tube. Stamen-filaments 1.5—2
cm. long, pilose at the base, adnate to the corolla tube up to c. 1 cm. from the

base; anther-thecae 2.3−3.5 mm. long, divergent. Disk 2 mm. long and 4−5 mm. in diam., cupuliform. Ovary 5−8 mm. long; style 17−27 mm. long. Capsule 20.84 × 1.8−2.5 cm., falciform to nearly straight, flattened, velvety tomentose with golden soft hairs, valves with a prominent longitudinally median ridge and two marginal ones near the dehiscing lines. Seeds 0.7−1.2 × 3.36 cm. including the wing.

Caprivi Strip: 1.6 km. outside Katima Mulilo on road to Ngoma, fl. 27.xi.1972, *Smith* 292 (K; LISC; SRGH). Botswana. N: Xlatseo, fl.12.xi.1974, *Smith* 1179 (K; SRGH). Zambia. B: Shangombo, 1020 m. fl. & fr. 8.viii.1952, *Codd* 7456 (BM; K; PRE). N: Mwewe-Mpundu, fl. 22.x.1949, *Bullock* 1349 (K; SRGH). W: Solwezi Distr., near Mutanda Bridge, fr. 2.vii.1930, *Milne-Redhead* 645 (K). C: Mt. Makulu Research Station 16 km. S. of Lusaka, fl. & fr. 5.ii.1957, *Simwanda* 93 (FHO, SRGH). E: 48 km. from Chipata on Rd. to Lundazi, fr. 26.iv.1952, *White* 2477 (K; FHO). S: Gwembe, Sinazeze village, fl. & fr. 4.xi.1955, *Bainbridge* 176 (K; FHO; SRGH). Zimbabwe. N: Sebungwe, near Binga-Kariyngwe Rd., 500 m. fl. ll.xi.1958, *Phipps* 1442 (K; SRGH). W: Hwange, Game Reserve 900 m. fl. & fr. 16.ii.1956, *Wild* 4738 (COI; K; LISC; SRGH). C: Chegutu Distr., 1200 m., st. iv.1927, *Jackson*. (SRGH). E: Mutare, 1080 m., fl. 18.i.1952, *Chase* 4330 (BM; SRGH). Malawi. N: Rumphi Boma, to Rumphi stream, fl. & immat. fr., 15.i.1953 *Chapman* 63 (FHO). C: Kasungu National Park, c. 1000 m., fl. 23.xii.1970, *Hall Martin* 1376 (PRE; SRGH). S: Zomba, fl. & immat. fr. i.1901, *Purves* 65 (K). Mozambique. N: Erati, Namapa, between C.I.C.A. Experimental Station and the lands Cabo Nauacha, road, Lúrio R., fl. & immat. fr. 8.iii.1960, *Lemos & Macuácua* 16 (COI; K; LISC; LMA; SRGH). Z: between Milange and Mocuba, fl. & fr. 11.iii.1943, *Torre* 4913 (C; LISC; LMA; MO). T: Chicoa, plateau of Serra Songo, c. 900 m. fl. 30.xii.1965, *Torre & Correia* 13932 (C; LISC; LMA; MO). MS: Manica, Dombe, near Matindire school, fl. 17.xi.1965, *Pareira & Marques* 681 (BM; COI; LISC; LMA; LMU; SRGH).

From Angola and Namibia to the east and from Zaire and Kenya to S. Africa. In open deciduous woodlands, xerophitic forests and secondary savannas, often on greyish sandy soils; 90−2400 m.

6. STEREOSPERMUM Cham.

Stereospermum Cham. in Linnaea, 7: 720 (1832).

Glabrous or pubescent trees or shrubs. Leaves opposite, imparipinnate. Panicles terminal, many-flowered large. Calyx campanulate or tubulate, irregularly 2−5 lobed. Corolla slightly zygomorphic, pink; tube infundibuliform, lobes patent. Stamens 4, didynamous inserted in the corolla tube, included; anther-thecae divergent; staminode 1, more or less developed. Disk entire or 5-lobed. Ovary linear-oblong, 4-angled, bilocular. Ovules numerous, 2-seriate per locule. Capsule longly linear-cylindrical. Seeds numerous, winged, arranged alternately along both sides of the septum.

A genus of c. 20 species centered in tropical Asia and extending with some 3−4 species into tropical Africa. In Flora Zambesiaca area only 1 species is known.

Stereospermum kunthianum Cham. in Linnaea, 7: 721 (1832).—Engl. & Prantl, Nat., Pflanzenfam. 4, 3b: 242 (1895).—K. Schum. in Engl. Pflanzenf. Ost-Afr. C: 364 (1895).—Sprague in F.T.A. 4, 2: 518 (1906).—Sillans in Notul. Syst. 14, 4: 325 (1952). —Pardy in Rhod. Agfric. Journ. 49: 81 cum 5 photogr. (1952). —Brenan in Mem. N.Y. Bot. Gard. 9: 18 (1954).—Williamson, Useful Pl. Nyasal.: 113 (1956).—Dale & Greenway, Kenya Trees & Shrubs: 66 (1961).—F. White, F.F.N.R.: 379 (1962).—Heine in F.W.T.A., ed 2, 2: 386 (1963).—Gomes e Sousa, Dendrol. Mocamb. 2: 665 (1967).— Paviani in Garcia de Orta 16: 173 (1968).—Drummond in Kirkia 10: 273 (1975).— Liben in Fl. Afr. Centr., Bignoniaceae: 14 (1977).—Palgrave, Trees of Southern Afr.: 832 (1981). TAB. 15. Type from Senegal.

Small or medium sized tree up to 15(20) m. or small shrub. Bark grey to whitish, smooth or flaking in plaques. Leaves 18−45 cm. long, 2−5-jugate; petiole 2.7−9(11) cm. long; leaflet-lamina 5−13(16) × 2−6(8.5) cm., ovate, obovate or elliptic, apex obtuse or acute or even apiculate, base cuneate or rounded and slightly asymmetric, margins entire or rarely toothed; superior surface glabrous, sometimes puberulous on the nerves, inferior surface minutely tomentose or velvety or glabrous, frequently with small glands on each side of the midrib near the base; petiolules very short or up to 1 cm. in the lateral leaflets even reaching 4.5 cm. in the terminal leaflet. Inflorescence a lax, terminal,

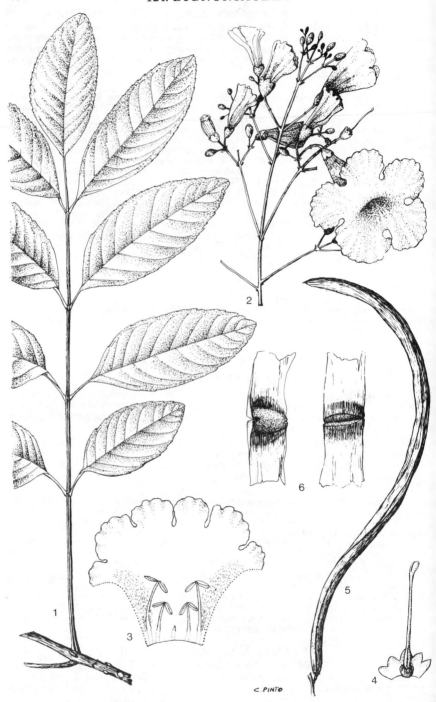

C. PINTO

Tab. 15. STEREOSPERMUM KUNTHIANUM. 1, part of branch with a leaf ($\times\frac{1}{2}$); 2, part of inflorescence ($\times\frac{1}{2}$); 3, corolla opened out showing stamens and staminode, viewed from inside ($\times\frac{3}{4}$), 1–4 from *Mendonça* 127; 5, capsule ($\times\frac{1}{2}$); 6, seeds, two views ($\times$1), 5–6 from *Simao* 5.

many-flowered panicle 15—48 cm. long, pubescent to tomentose or glabrescent; flowers appearing before the leaves; pedicels 4—12 mm. long; bracts 3—4 mm. long, oblong. Calyx 5—8(10) mm. long campanulate, slightly 5-lobed, or almost truncate tomentose outside mainly on the lower half. Corolla funnel shaped, with a pinkish tube 30—40 mm. long, showing red streaks on the throat and along the lower corolla lobes, puberulous outside, glabrous inside except in a band opposite to staminode; corolla lobes c. 20 mm. in diam. more or less circular, lighter coloured than the tube. Stamen-filaments 6—15 mm. long; anthers 2 mm. long; staminode 1.5 mm. long. Disk 1.5 mm. long. Ovary 5—6 mm. long, glabrous, reddish-brown. Seeds 7 × 20—30 mm. including the wing.

Zambia. B: Kabompo R., fr. 1938, *Martin* 900 (FHO). N: Bulaya, Chishela Dambo, NE. of Mweru-wa-Ntipa, 943 m. fl. 11.viii.1962, *Tyrer* 402 (BM; SRGH). W: Ndola, fl.viii.1933, *Duff* 157 (K; FHO). C: Mt. Makulu Research Station 16 km. S. of Lusaka, fl. 12.ix.1956, *Kassam* 6 (K; FHO; SRGH). E: Imsefula R. to Nyimba R., fl. 25.viii.1929, *Burtt Davy* s.n. (FHO). S: Choma, Mapanza, 1060 m. fl. 13.ix.1958, *Robinson* 2885 (K; PRE). **Zimbabwe.** N: Lomagundi Distr., Banket, Colburnie Farm, fl. & fr. 14.x.1948, *Armitage* 10 (COI; K; SRGH). W: Hwange Distr., Zmabezi R., c. 3 km. on Matetsi R. side of Sidenda Isl., fl. 16.viii.1971, *Lovemore* 588 (K; SRGH). C: Harare 1500 m., fl. 10.x.1924, *Eyles* 1137 (K). E: Mutare Distr., fl. 11.x.1949, *Chase* 1690 (BM; COI; K; LISC; SRGH). **Malawi.** N: Nkhata Bay, Chikale beach, 460 m., fl. & immat fr. 8.xi.1975, *Pawek* 10344 (K; Ma; MO; SRGH; UC). C: W. Lake Malawi Hotel, 320 m. fl. 4.viii.1951, *Chase* 3889 (BM; K; LISC; SRGH). S: Chikwawa, 300 m., fl. 5.x.1946, *Brass* 17928 (K; SRGH). **Mozambique.** N: Mutuali, near the Catholic Mission, fl. 2.ix.1953, *Gomes e Sousa* 4105 (COI; K; LISC; LMA; SRGH). Z: Mocuba, Namagoa Estate, 120 m., fl. x, *Faulkner* 463 (COI, K; PRE; SRGH). T: Cahora Bassa, Mecangadzi R., fr. 17.x.1973, *Correia, Marques & Pereira* 3465 (K; LISC; LMU; n.v.; SRGH). MS: Sofala, Cheringoma, between Inhaminga and Lacerdonia, fl. 13.vii.1941, *Torre* 3088 (BR; COI; FHO; K; LISC; LUA).

From Senegal to Ethiopia and southwards to Zimbabwe and Mozambique. In open deciduous forests, sandy savanna woodlands, *Brachystegia* or *Adansonia-Cordyla* woodlands and riverine forests; 60—1500 m.

7. PODRANEA Sprague

Podranea Sprague in F.C. 4, 2: 449 (1904).

Shrubs or subshrubs, almost glabrous. Leaves imparipinnate, opposite. Panicles terminal with pinkish or lilac flowers. Calyx campanulate, regularly 5-lobed, tube inflated, opening early in bud. Corolla slightly bilabiate tube cylindrical widening at the top. Stamens 4, didynamous, adnate to the corolla tube, anther-thecae connate only at the apex, divaricate. Staminode 1. Disk annular-cupular. Ovary oblong, bilocular. Ovules 8-seriate in each locule. Capsule, slender and flattened. Seeds many, winged.

A genus with only 2 species, both African.

Podranea brycei (N.E. Br.) Sprague in F.T.A. 4, 2: 5154 (1906).—Dyer in Fl. Pl. Afr. 34: t. 1348 (1961).—Drummond in Kirkia 10: 273 (1975). TAB. 16. Type: Zimbabwe, Mashonaland, in dry places, 1350 m., fl. v.1896, *Bryce* s.n. (K, holotype). *Tecoma brycei* N.E. Br. in Kew Bull. 1901: 130 (1901). Type as above.

Scandent shrubs up to 10 m. high or subshrubs. Branches 4-angled, grooved minutely lepidote, pubescent only at the nodes. Leaves imparipinnate, 13—27 cm. long, 4—7-jugate; petiole 2—6 cm. long; leaflet lamina 3.5—9 × 1.3—2.6 cm., narrowly ovate to ovate-lanceolate, longly acuminate at the apex, rounded to somewhat asymmetric at the base, grass green, inferior surface slightly lighter and glandular, margins entire or sometimes serrate-crenate uni- or bilaterally along the distal half; petiolules up to 1.5 cm. long in proximal pairs, decreasing progressively upwards. Inflorescence a few- to many-flowered terminal panicle, up to 40 mcm. long, glabrous, branches opposite and decussate; pedicels 1—2 cm. long, articulated c. 3.5 mm. from the top, upper articulate, falling off with the flower, the lower persistent and 2-bracteate towards the base. Calyx whitish, campanulate, 5-ribbed, inflated, 5-lobed, lobes triangular-oblong, 0.6—1.2 × 0.4—0.7 cm., apiculate at the apex and revolute, tube 0.7—1.5 cm. long, glabrous outside and minutely glandular inside. Corolla campanulate, mauvish pink, throat brighter and carmine streaked inside; corolla tube 3.5—5 cm. long,

Tab. 16. PODRANEA BRYCEI. 1, flowering branchlet (×½); 2, corolla opened out showing
stamens and staminode, viewed from inside and the gynoecium (×½); 3, stamen, dorsal
view (×1); 4, stamen, ventral view (×1), 1−4 from *Macedo* 2200; 5, capsule dehiscing
(×½); 6, seed (×1), 5−6 from *Noel* 3859.

cylindrical at the base c. 1 cm. long suddenly campanulate near the top, glabrous outside and villous inside, limb 5-lobed, bilabiate, lobes 1.3−2.8 cm. diam., almost circular, margin ciliate and pilose at the sinus. Stamens didynamous, included, with the staminode between the shorter pair; filaments adnate up to 1 cm. from the base of the corolla tube, densely pilose at the insertion points; anther-thecae divergent when mature, the connective apiculated and dorsaly crested. Disk c. 1 mm. long, annular-cupular. Ovary c. 6 mm. long. Capsule 35−48 × 1.5 cm., apex apiculate. Seeds 0.6−1 × 3 cm. including the wings.

Zimbabwe. N: Mazoe Distr., fl. 10.vi.1946, *Wild* 1118 (K; LISC; SRGH). W: Bulawayo, 1350 m., fl. iv.1921, *Borle* 155 (K). C: Marondera fl. 9.x.1942, *Dehn* 353 (K; SRGH). E: Chimanimani, fl. 7.vii.1950, *Crook* M61 (K; LISC; LMU; SRGH). S: Lundi R., fl. & fr. 30.vi.1930, *Hutchinson & Gillett* 3307 (BM; K; LISC; SRGH). **Malawi.** S: Nsanje Distr., Matandwe Forest Reserve, near Mbeu Resthouse, 750 m., 9. viii. 1960, *Willan* 44 (BM; FHO; LISC; SRGH). **Mozambique.** N: Unango, fl. 27. v. 1948, *Pedro & Pedrogão* 3955 (LMA). MS: Manica, Chimoio, Garuzo, Fl. 2.iii.1948, *Andrada* 1089 (COI; FHO; LISC; LMA; SRGH).
Confined to the Flora Zambesiaca area. In mixed deciduous forests and along watercourses; c. 750-1500 m.

8. FERNANDOA Welw. ex Seem.

Fernandoa Welw. ex Seem. in Journ. Bot. Lond. **3**: 330, 333, t. 37−38 (1865) *"Ferdinandia"*, corr. Seem. op. cit. **4**: 123 (1866); op. cit. **8**: 280 & 403 (1870) *"Ferdinandia"*, corr. Seem. op. cit. **9**: 81 (1871).−K. Schum. in Engl. & Prantl, Nat. Pflanzenfam. 4, **3b**: 243 (1895).−Heine in Adansonia N.S. **4**: 467 (1964).− Milne-Redhead in Kew Bull. **1948**: 170 (1948).

Shrubs or trees. Leaves caducous, imparipinnate, opposite, Flowers large, in short few-flowered racemes or solitary, borne praecociously and in axils of fallen leaves. Calyx campanulate irregularly lobed, opening just before the corolla anthesis. Corolla broadly cup-shaped, tube short, limb slightly bilabiate. Stamens 4, slightly exserted; staminode 1; anther-thecae divaricate. Disk cupular. Ovary cylindrical, narrow, bilocular; ovules pluriseriate in each locule. Capsule subcylindrical, dehiscing at right angles to the thin flat septum. Seeds numerous, winged.

An African genus with 3 continental species and 3 other in Madagascar (originally described as species of the genera *Colea* Bojer and *Kigelia* DC. and which Gentry in Ann. Miss. Bot. Gard. **62**: 48 (1975) has transferred to *Fernandoa*, viz. *F. coccinea* (Scott Elliot) Gentry, *F. macrantha* (Baker) Gentry and *F. madagascariensis* (Baker) Gentry, which the 3 had been considered until then in the Malagasian genus *Kigelianthe* Baill. which is not accepted.

Fernandoa magnificia Seem. in Journ. Bot. **8**: 280 & 403 (1870) *"Ferdinandoa"*, corr.− Seem. op. cit. **9**: 81 (1871).−Spragus in F.T.A. **4**, 2: 517 (1906) *"Ferdinandia"*.−Milne-Redhead in Kew Bull. **1948**: 171 (1948).−Heine in Adansonia, N.S. **4**: 470 (1964).− Gomes e Sousa, Dendrol. Moçamb. **2**: 668 (1967).−Paviani in Garcia de Orta **16**: 175 (1968).−Drummond in Kirkia **10**: 273 (1975).−Palgrave, Trees of Southern Afr.: 832 (1981). TAB. **17**. Type from Tanzania.
Heterophragma longipes Baker in Kew Bull. **1894**: 31 (1894). Syntypes from Tanzania.

Shrub, simple or many-stemmed from the base, or small tree 2−7.7(12) m. tall (reaching 30 m. outside Flora Zambesiaca area). Bark brown or greywish, glabrous, rough, longitudinally deeply fissured with prominent elliptic, yellow-whitish lenticles. Leaves large, imparipinnate, (4)5−7-jugate; rhachis 5−25 cm.; petiole (1)2.5−9 cm. long, leaflet laminae 3−14(16) × 2−6 cm., lanceolate, ovate to oblong, apex attenuate to longly acuminate, base broadly cuneate to rounded and more or less asymmetric, margins entire or almost so, not so often irregularly toothed to crenulate, undulate with conspicuous reticule on both surfaces, nerves impressed on superior surface, prominent on the inferior one and provided with axillary pubescent domatiae; petiolules of lateral leaflets sessile or almost, of the terminal one reaching 2.8 cm. Inflorescence an axillary racemose cyme, 13−22 cm. long few-flowered, or flowers solitary, produced before the leaves of the year; pedicels 5−7 (10) cm. long, walking-cane shaped.

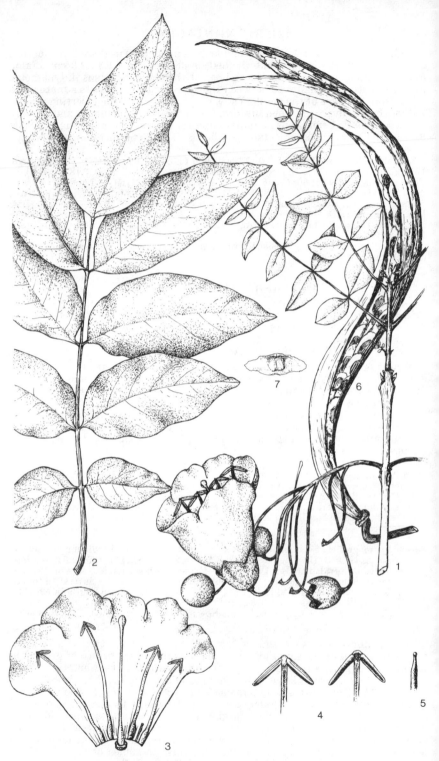

Tab. 17. FERNANDOA MAGNIFICA. 1, flowering branchlet (×½), *Mendonça* 2003; 2, leaf (×½), *Müller & Pope* 1903; 3, corolla opened out showing stamens and staminode, viewed from inside and the gynoecium (×½); 4, stamens (×1); 5, staminode (×5), 3—5 from *Mendonça* 2003; 6, capsule showing septicidal dehiscence (×½); 7, seed (×½), 6—7 from *Simão* 653/48.

Calyx 1.3−2.3 cm. long, bell-shaped, irregularly 3−5 lobed halfway down, lobes ovate-deltoid, mucronulate, with scattered small glands. Corolla 5−9 cm. long, broadly campanulate slightly 2-labiate, lobes more or less rounded and unequal, orange-reddish or crimson, yellow winish streaked at the throat. Stamens 4, filaments flattened and adnate up to c. 8 mm.; anthers (6)8−11 mm. long, reaching the mouth of the corolla tube, thecae divergent, connective beaked. Ovary 6−11 (18) mm. long cylindrical, glabrous; ovules 4−6-seriate in each locule; style 4−6 cm. long, flattened. Capsule terete, 33−54 × 1.2 cm., spirally twisted and curved. Seeds 7−12 × 18−36 mm. (including wings).

Zimbabwe. E: Chipinge, lower Sabi valley, fl.ix.1960, *Soane* 310 (K; SRGH). **Malawi**. S: Nsanje, fr. 5.i.1964, *Chapman* 2191 (FHO, SRGH). **Mozambique**. N: Macondes, between Nangade and Mueda, fl. & fr. 18.ix.1948, *Barbosa* 2207 (COI; K; LISC; LMA). Z: Between Nante and Maganja da Costa, 60 m., fl. 17.ix.1964, *Gomes e Sousa* 4830 (COI; K; LMA; SRGH). MS: c. 8 km. S. of Buzi Ferry, fl. 28.viii.1961, *Leach* 11235 (BM; COI; K; SRGH). GI: Govvuro, 70 km. from Mabote to Jofano, fl. 12.ix.1973, *Correia & Marques* 3347 (LISC; LMU).
Also in Kenya and Tanzania. In dense mixed dry forests, edges or evergreen forests; from 60 m.

9. KIGELIA DC.

Kigelia DC. in Bibl. Univ. Genève, Sér. 2, **17**: 135 (1838).

Medium or large tree with a rounded crown. Leaves imparipinnate, opposite or ternate. Inflorescence a terminal, lax, pendulous panicle, longly pedunculate. Calyx campanulate to almost tubular, irregularly lobed, coriaceous. Corolla large, widely campanulate; limb bilabiate and 5-lobed; corolla tube cylindrical at the base, widening and uncurving upwards. Stamens 4, didynamous, adnate to the top of the cylindrical part of the corolla tube; staminode 1, rather large. Disk annular, thick. Ovary terete, unilocular, placenta parietal 2, widely intruded; ovules numerous, pluriseriate. Fruit large, cylindrical, pendulous, wood-walled, indehiscent, with a fibrous pulp. Seeds numerous, thick; wingless, testa coriaceous.

A tropical African monospecific genus.

Kigelia africana (Lam.) Benth. in Hook., Niger Fl.: 463 (1849).−Sprague in F.T.A. **4**, 2: 536 (1906).−Heine in F.W.T.A., ed. 2, **2**: 385 (1963).−Merxm. & Schreiber in Merxm., Prodr. Fl. Sw. Afr. **128**: 3 (1967).−Paviani in Garcia de Orta **16**: 175 (1968).−Palmer & Pitman, Trees of Southern Afr. **3**: 2011 cum 2 photogr. & 2 fig. (1973).−Drummond in Kirkia **10**: 273 (1975).−Compton, Fl. Swazil.: 538 (1976).−Liben in Fl. Afr. Centr., Bignoniaceae: 4, t. 1 (1977).−Palgrave, Trees of Southern Afr.: 833 (1981). TAB. 18. Type from Senegal.
Bignonia africa Lam., Encycl. Méth., Bot. **1**: 424 (1785). Type as above.
Crescentia pinnata Jacq., Collect. **3**: 203, t. 18 (1789). Type ... ??
Tanaecium pinnatum Willd. in L., Sp. Pl. ed. 4, **3**: 312 (1800). Type as for *Crescentia pinnata*.
Kigelia pinnata (Jacq.) DC., Prodr. **9**: 247 (1845).−Klotzsch in Peters, Reise Mossamb., Bot. **1**: 195 (1861).−Pardy in Rhod. Agric. Journ. **50**: 3656 cum 3 photogr. (1953). −Williamson, Useful Pl. Nyasal: 73 (1956).−F. White, F.F.N.R.: 379 (1962).−Gomes e Sousa, Dendrol. Moçamb. **2**: 662 (1967). Type as for *Crescentia pinnata*.
Kigelia aethiopica Decne in Deless., Ic. Sel. Pl. **5**: 39, t. 93A e 93B (1849).−Schinz in Denkschr. Math.-Naturwiss. K. Kais. Akad. Wiss. **78**: 439 (1905).−Williamson, Useful Pl. Nyasal.: 72 (1956). Type from Ethiopia.
Kigelioa pinnata var. *tomentella* Sprague in F.T.A. **4**, 2: 537 (1906). Syntypes from Botswana and Zimbabwe: Zambesi R., Victoria Falls, fl., s.d. *Allen* 30 (K, lectotype here designated; LISC and SRGH, photolectotype); paratypes: Zimbabwe, Isl. in Zambesi R., 9.6 km. above Victoria Falls *Cartwright* s.n. (... , not seen); Botswana, Ngamiland, Tamala Kane R., *McCabe* 45 (... , not seen); Okavango valley, Bakalahari village, *Lugard* 233 (... , not seen).

Medium or large sized tree up to 25 m. tall. Leaves opposite or in whorls of 3, imparipinnate, crowded towards the tops of the branches; leaflets (1) 2−5-jugate, sessile or subsessile, except the terminal ones with petiolule (0.7)1−4(6.5) cm., long; leaflet lamina 3.5−17.5 (22.5) × 2.5−11 cm., ovate elliptic, obovate to rounded, apex obtuse, broadly tapering to rounded or retuse and not so often apiculate, base rounded to cuneate, slightly to profoundly asymmetric except in the terminal leaflet which is asymmetric, glabrous to more or less hairy in both surfaces sometimes more roughly hairy in the superior one, papyraceous to coriaceous, margins entire, serrate or toothed and sometimes conspicuously wavy; lateral nerves (4)6−13 pairs impressed above, prominent below, venation laxly reticulate; petiole (2)3.5−14(16) cm. long; rhachis 3−25 (29) cm. long, sulcate above, terete below. Flowers in pendulous very lax, terminal panicles, 30−100 (150) cm. long, longly pedunculate; pedicels 1−11 (13.5) cm. long, upcurved at the tip; bracts small lanceolate caducous. Calyx shortly tubular to campanulate, (1.7)2−4.3 cm. long, irregularly 4−5 lobed with the lobes up to 1 cm. long, ribbed, glabrous to sparsely puberlous outside, sometimes with irregularly scattered small glands. Corolla large, 6−12 cm. long, widely cup-shaped, at first yellowish, later becoming reddish to purplish, streaked darker inside and outside, glabrous except sometimes at the point where the filaments become free; limb bilabiate, the superior lip bilobed, the lower one 3-lobed and recurved, lobes more or less rounded; corolla tube cylindrical at the base and suddenly widening and incurving upwards. Stamen-filament 3.5−6.5 cm. long, adnate up to 1−2.5 cm. from the base of the corolla tube; anthers 7−13 mm. long; staminode rather large. Disk c. 1 cm. in diam. 2−3 mm. high, fleshy, irregularly lobed, sometimes almost truncate. Ovary 8−15 mm. long, cylindrical; style 4−7(8) cm. long, filiform. Fruit sausage-shaped up to 1 m. long and 18 cm. in diam., pendulous from a long peduncle, greyish-brown, lenticellate in the youth, massive, wood-walled, indehiscent. Seeds 10 × 7 mm. numerous, wingless, embedded in a fibrous pulp; testa coriaceous; cotyledons folded.

Botswana. N: Mbone Isl., fl. 17.x.1974, *Smith* 1158 (K; SRGH). Zambia. B: near Senanga, near lake, 1020 m., fl. 3.viii.1952, *Codd* 7381 (BM; K; PRE; SRGH). N: Mbala Distr., 1450−1800 m., fl. ix. 1935, *Gamwell* 243 (BM). W: Solwezi Distr., near Lunsala Bridge, fr. 2.vii.1930, *Milne-Redhead* 650 (K). C: Makulu Mt., Research Station, 16.09 km. S. of Lusaka, fl. 14.ix.1960, *Coxe* 1 (FHO; K; SRGH). E: Petauke Distr., E. bank of the Luangwa R., above the Beit Bridge, fl. 5.ix.1947, *Brenan & Greenway* 7806 (FHO; K). S: Pemba, 1050 m., fl. ix.1909, *Rogers* 8554 (K; PRE). Zimbabwe. N: Mazoe, fl. viii.1910, *Bell* 929 (BM; SRGH). W: near Victoria Falls, banks Zambesi R., 870 m. fl. & immat. fr. 11.ix.1905, *Galpin* s.n. (ORE). C: Harare in Greenwood Park, fl.viii.1937, *McGregor* 109 (BM; FHO; SRGH). E: Odzi Distr., Dice Box Farm, Odzi R., fl.22.viii.1949, *Chase* 1805 (K; SRGH). S: Lomagundi Distr., Plateau Farm, Mhangura area, fl. 17.x.1964, *Jacobsen* 2497 (PRE). Malawi. N. Nyungwe, near Lake, fl. 13.ix.1930, *Migeod* 922 (BM). C: Nkhota Kota Distr., Chia area, 480 ml. fl. 7.ix.1946, *Brass* 17557 (BM; K; SRGH). S: Chikwawa Distr., lower Muanza R., 180 m. fl. 4.x.1946, *Brass* 17947 (K; SRGH). Mozambique. N: Mecufi, near the old Administrative post of the Lúrio, fl. & fr. 21.viii.1948, *Andrada* 1289 (COI; EA; LISC; LMA; LMU). Z: Mocuba, Namagoa, fl. & immat. fr. viii.1949, *Faulkner* 307 (K; LMA; PRE; SRGH). T: Chiôco, c. 27 km. from Chiôco, fr. 2.xi.1965, *Myre & Rosa* 4764 (LISC., LMA). MS: Sofala, Chupanga, fl.viii.1858, *Kirk* s.n. (K). GI: Gaza, Caniçado, 10 km. from Mabalane (Vila Pinto Teixeira) to Balula on the road along the Limpopo R., fl. 25.viii.1969, *Correia & Marques* 1219 (COI; LISC; LMA; SRGH). M: Maputo, near the Polana beach, fl. 14.xii.1940, *Torre* 2424 (C; LISC; LMA; MO; WAG).

Widespread in tropical Africa. In rain forests, open woodlands, fringing woodlands, xerophitic forests, or savanas; usually on sandy argillaceous soils; 0−2000 m.

This species shows great variability in habit and foliar morphology, and this led Sprague to create 10 species and 4 varieties (F.T.A. 4, 2: 533, 1906). *Aubreville* and *Sillans* recognised later (cf. Sillans in Not. Syst. 14: 323, 1953) only one species with 3 varieties: *Kigelia africana* var. *africana*, var. *aethiopica* (Decne.) Aubrev. ex Sillans and var. *elliptica* (Sprague) Sillans, for W. Africa. Although we recognise that the specimens growing in woodlands and in forests present a tendency to produce large leaflets with acute apices, entire margins, dense indumentum and weak consistence, as opposed to the tendency in specimens from open areas, to produce smaller leaflets, with rounded apices, toothed margins, glabrescent and coriaceous consistence. We noticed all steps of variability to occur, and thus believe that they are mere ecoforms.

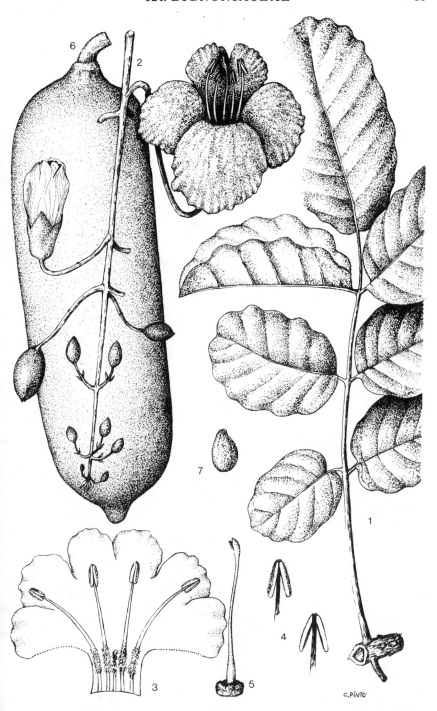

Tab. 18. KIGELIA AFRICANA. 1, leaf (×½), *Andrada* 1289; 2, part of inflorescence (×½), *Codd* 7381; 3, corolla opened out showing stamens and staminode (×½); 4, dorsal and ventral side of stamen (×1); 5, gynoecium (×½); 3—5 from *Torre* 2424; 6, small mature fruit drawn from various sources (×¾); 7, seed (×½), from *Andrada* 1289.

125. PEDALIACEAE

By H.-D. Ihlenfeldt

Small trees with swollen stems, shrubs with or without swollen main branches, perennial herbs, sometimes with a short swollen stem and tuberous roots, or annual herbs, erect or procumbent, covered with mucilage-glands (at least on the young parts) which produce slime when wetted. Leaves opposite or the upper sometimes alternate, petiolate to subsessile, usually simple, entire to pinnatilobed, sometimes digitate, exstipulate, sometimes subsucculent. Flowers usually solitary in the leaf axils, rarely in few-flowered cymes; pedicels generally with nectar glands (extrafloral nectaries originating from reduced flowers) at the base. Flowers hermaphrodite, irregular. Calyx 5-partite. Corolla gamopetalous; tube usually obliquely campanulate, sometimes funnel-shaped or cylindrical, adaxially often slightly gibbous or rarely spurred at the base; limb sub-bilabiate or subequally 5-lobed. Stamens 4, didynamous (fifth stamen often represented by a staminode), usually inserted near the base of the corolla and normally included in the tube; thecae 2, parallel or divaricate, opening lengthwise; connective usually gland-tipped. Disk (nectary) hypogynous, fleshy, generally conspicuous, often asymmetrical. Ovary superior, usually bilocular, the loculi often completely or partially divided by false septa, each compartment containing 1-many ovules attached to a central placenta; style filiform, exceeding the anthers; stigma usually bilobed. Fruit very variable, dehiscent or indehiscent, often provided with protuberances such as horns, spines or wings. Seeds 1-many in each compartment; testa often characteristically sculptured, sometimes forming wings; seeds containing considerable amounts of fat; the endosperm very thin.

A family of 13 genera with approximately 60 species, native of Africa, extending to India and Sri Lanka (*Sesamum* L.) and Australia (*Josephinia* Vent.); one genus (*Uncarina* (Baill.) Stapf) endemic to Madagascar. *Sesamum indicum* L. is an important crop plant, cultivated in most tropical and subtropical countries of the world for the oil which is extracted from the seeds.

1. Spiny shrub with trunk and main branches swollen, 1—4 m. tall; leaves deciduous, alternate on long shoots, fasciculate on short shoots in the axils of the spines; flowers in few-flowered raceme-like inflorescences; corolla with a long cylindrical tube, slightly curved, produced near the base into 1—1.5 cm. long blunt spur; fruit a rigid woody capsule, laterally compressed, bilocular, each loculus divided by an incomplete false septum; seeds flat with broad membranous wings, up to 2.5 cm. in diam. (wings incl.) (tab. 19) - - - - - - - - - - - 1. **Sesamothamnus**
   − Perennial herb, sometimes with a short swollen stem (if so, plant not more than 50 cm. high), or annual, erect, ascending or procumbent; leaves never fasciculate in the axils of spines; flowers axillary, solitary, very rarely in groups of 3, tube not spurred (except in *Holubia*), but sometimes slightly gibbous at the base; fruit variable but never a laterally compressed woody capsule - - - - - - 2
2. Fruit with a 1—20 mm. broad parchment-like wing on each of the four edges, laterally compressed, more or less rectangular in cross section - - - - 3
   − Fruit without parchment-like wings - - - - - - - 4
3. Annual erect herb; corolla cream, at the base with a large, sac-like spur; leaves circular-ovate, slightly lobed or sinuate; fruit up to 60 mm. long and 50 mm. broad, with wings up to 20 mm. broad; the veins of the wings forming a dark network (tab. 20) - - - - - - - - - - - - - 2. **Holubia**
   − Perennial herb with a persistent short stem and a tuberous root or a swollen tuber-like, partially subterranean stem up to 15 cm. high from which annual erect branches arise; leaves linear to lorate, sinuate to pinnatilobed; corolla yellowish to wine red, without a basal spur (but sometimes slightly gibbous at the base); fruit up to 35 mm. long and 41 mm. broad, the veins of the wings indistinct (tab. 21) - - - - - - - - - - - - - - - 3. **Pterodiscus**
4. Fruit a dehiscent non-woody capsule; erect annual or perennial herb - - 5
   − Fruit woody, either indehiscent or tardily dehiscent only at the apex; armed with spines or hooks; procumbent or ascending annual or perennial herb - - 6

5. Capsule obtuse or truncate at the apex, with two lateral horns at the angles of the apex (sometimes not very conspicuous); seeds never winged; largest leaves usually sagittate or 3-lobed (tab. 22) - - - - - - - **4. Ceratotheca**
— Capsule without lateral horns at the apex; seeds sometimes with conspicuous wings; leaves very variable (tab. 23) - - - - - - - **5. Sesamum**
6. Fruit disk-like with two erect conical spines from near the centre; flowers on long slender pedicels, corolla obliquely campanulate, sub-bilabiate, white or pink, the tube inside streaked with a darker colour (tab. 26) - - - **6. Dicerocaryum**
— Fruit not disk-like; flowers sessile or on shortly pedicellate, usually purple or yellow, corolla tube either funnel-shaped or cylindrical and constricted at the base, corolla lobes nearly equal - - - - - - - - - - - 7
7. Fruit subpyramidal, quadrangular with a spine (sometimes rather inconspicuous) at each basal angle; flowers yellow, corolla tube funnel-shaped; ascending annual herb (tab. 27) - - - - - - - - - - - **7. Pedalium**
— Fruit laterally compressed, ovate or oblong in lateral view, with two obtuse protruberances on each face, and in addition either with two rows of curved arms along both edges, each bearing recurved spines, or edges with two rigid wings bearing recurved spines; flowers usually purple (but tube often yellowish), corolla tube cylindrical, constricted at the base; procumbent perennial herb with tuberous roots (tab. 28) - - - - - - - - - - **8. Harpagophytum**

1. SESAMOTHAMNUS Welw.

Sesamothamnus Welw. in Trans. Linn. Soc. **27**: 49, t. 18 (1869).
Stigmatosiphon Engl. in Bot. Jahrb. **19**: 150 (1894).—Bruce in Kew Bull. **8**: 417 (1953).

Branched, spiny, small trees or shrubs; trunks smooth, usually swollen at the base, with ascending branches. Leaves deciduous, entire, usually obovate, petiolate and alternate on long shoots, almost sessile and fasciculate in the axils of spines (modified petioles). Flowers large, white, cream, pink or (not in the Flora Zambesiaca area) yellow, sweet-scented in the evening and early morning (pollinated by butterflies), in few-flowered raceme-like inflorescences. Calyx 5-partite, subequally lobed, posterior lobe smaller. Corolla with a long cylindrical or (not in the Flora Zambesiaca area) narrowly funnel-shaped tube, curved or (not in the Flora Zambesiaca area) straight, usually with a conspicuous spur near the base; limb spreading at right angles to the tube, lobes subequal, entire or (not in the Flora Zambesiaca area) fringed. Stamens 4, subequal, inserted near the throat of the corolla tube; thecae parallel; pollen in tetrads. Ovary bilocular divided incompletely by false septa into 4 compartments; ovules numerous, uniseriate in each compartment. Fruit a rigid woody capsule, laterally compressed, oblong to obovate in outline. Seeds numerous, large, compressed, subcircular to transversely oblong, with broad membranous wings.

A genus comprising 6 species, all native of Africa.

Sesamothamnus lugardii N.E. Br. ex Stapf in F.T.A. **4**, 2: 568.—Bruce in Kew Bull. **8**: 417 (1953).—Codd, Trees & Shrubs Kruger Nat. Park: 168 (1951).—Pardy in Rhod. Agric. Journ. **53**: 63 (1956).—Codd in Fl. Pl. Afr. pl. 1640 (1972).—Palmer & Pitman, Trees of S. Afr.: 2015 (1972).—
Coates & Palgrave, Trees of S. Afr.: 835 (1977). TAB. **19**. Type: Botswana, Northern Kalahari desert, near Tsokotse Salt Pan, 1500 m., *Lugard* 274 (K, holotype).
Sesamothamnus seineri Engl., Pflanzenw. Afr. **1**: 586, t. 28, 2 (1910), nom. nud.

Soft-stemmed shrub up to 4 m. tall, trunk swollen at the base, up to 1 m. in diam., from which arise several thick, erect main branches which taper rapidly; bark grey or mottled with yellow; branches stiff, sparingly branched, spiny; spines spreading to recurved, 5—10 mm., leaves deciduous, oblong to obovate, 1—2.5 cm. long, 4—6 mm. broad, obtuse to retuse at the apex, cuneate at the base, grey, semi-coriaceous, densely covered with stellately-branched mucilage glands; midrib distinct below, impressed above. Flowers large, sweet-scented, 1—3 in short raceme-like inflorescences on short shoots which are sometimes elongated; pedicels 4 mm. long. Calyx 5-partite, 4 mm. long. Corolla sparingly tomentose, with a long, cylindrical, only slightly curved tube, 8—10 cm. long, 4—6 mm. in diam., produced near the base into a short blunt spur, 1—1.5 cm. long, cream-coloured, sometimes suffused with purple; limb spreading at first, later reflexed, 4—5 cm. in diam., slightly oblique; lobes 5, subcircular, 1.6—2

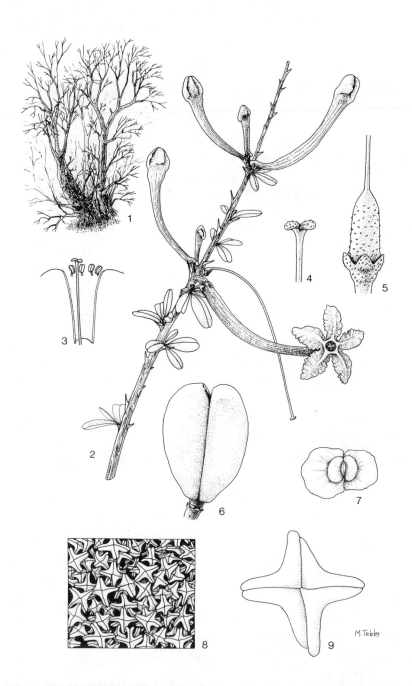

Tab. 19. SESAMOTHAMNUS LUGARDII. 1, habit (×$\frac{1}{30}$); 2, branch with flowers (×1); 3, corolla opened to show stamens and stigma (×1); 4, stigma (×3); 5, calyx and ovary (×3); 1−5 from Fl. Pl. Afr. pl. 1640 (1972); 6, fruit in lateral view (×1); 7, seed (×1); 8, portion of inferior leaf surface with mucilage-glands (×75); 9, single mucilage-gland (×300); 6−9 from *Ihlenfeldt* 2125.

cm. broad. Stamens 4, subequal, inserted near the throat of the corolla tube; filaments 4—5 mm. long. Ovary, oblong, 7—8 mm. long, laterally compressed, bilocular, each loculus divided by an incomplete false septum; ovules many, uniseriate in each compartment. Style slender, equaling the corolla tube; stigma reaching the corolla mouth or exerted by up to 10 mm., broadly bilobed. Fruit a rigid woody capsule, obovate in outline, often retuse at the apex, 4—6 cm. long, 3.5—5 cm. broad, laterally compressed, greyish brown. Seeds flat, transversely oblong, winged, about 1.5 cm. long and 2.5 cm. broad (incl. wings).

Botswana. N: Francistown, fl. iv.1926, *Rand* s.n. (BM). SE: Between Lake Xau and Tsokotse Pan, fr. 13.iii.1965, *Wild & Drummond* 7239 (K; PRE; SRGH). **Zimbabwe.** W: Hwange, Gwai Bridge, Tsholotsho Rd., *Orpen* 35/53, fl. x.1953 (K; SRGH). S: Flood plain N. of Shashi-Limpopo confluence, fr. 22.iii.1959, *Drummond* 5935 (K; SRGH).

Also in S. Africa. On calcareous soil or between rocks, in scrub communities, sometimes being the dominant woody plant.

2. HOLUBIA Oliver

Holubia Oliver in Hooker, Ic. Pl. t. 1475 (1884).

Erect annual (sometimes biennial) herb with spreading branches, 30—75 cm. tall with square stem. Leaves opposite, long petiolate; lamina subsucculent, circular to ovate, slightly lobed or sinuate, truncate or slightly cordate at the base, sparingly glandular below, glabrous above. Flowers solitary, yellow-green to white; corolla tube cylindrical at the middle, funnel-shaped at the mouth, with a large basal sac-like spur; limb spreading, obscurely bilabiate. Stamens 4, inserted low down in the corolla tube; thecae divergent. Ovary bilocular, loculi undivided; ovules biseriate in each loculi, c. 8 in each loculi, ascending. Fruit indehiscent, more or less rectangular in cross section, with a broad parchment-like wing on each of the four edges, rotund to circular in lateral view; mesocarp not spongy; the veins of the wing forming a distinct, usually dark coloured network; c. 8 seeds in each loculus.

A monotypic genus, endemic to southern Africa.

Holubia saccata Oliver in Hooker, Ic. Pl. t. 1475 (1884).—Stapf in F.C. 4, 2: 457 (1904); in F.T.A. 4, 2: 547 (1906).—Bolus in Ann. Bolus Herb. 1: 133, t. 1.—Ihlenf. in Mitt. Staatsinst. Allg. Bot. Hamburg 12: 61 (1967), t. 2. Tab. 20. Type from S. Africa.

Erect annual (sometimes biennial) herb with spreading branches, 30—75 cm. tall. Leaves opposite, petiolate; lamina subsucculent, circular to ovate, slightly lobed or sinuate; up to 7 cm. long and broad. petiole 2.5—7 cm. long; Flowers solitary, yellow-green, yellow, cream or white, with unpleasant odour; corolla tube (spur excluded) 30—40 mm. long, limb up to 50 mm. in diam.; spur up to 30 mm. long and 20 mm. wide. Fruit rotund to circular in lateral view, up to 60 mm. long and 50 mm. broad, often suffused with purple. Seeds dark brown, obovate in outline, c. 6 mm. long and 4 mm. broad; testa with reticulate sculpturing.

Botswana. N: Francistown, fl. & fr. 7.iii.1961, *Richards* 14544 (K; SRGH). SE: Tlalamabele/Mesu area near Boa Pan, fl. 13.i.1974, *Kockott* 278 (SRGH). **Zimbabwe.** W: Bulawayo, Antelope Mine, fl. 23.iii.1950, *Orpen* 061/50 (SRGH). E: Nyanyadzi, Sabi Valley, fr. iii.1972, *Goldsmith* 42/72 (SRGH). S: Beit Bridge, fl. & fr. 10.i.1961, *Leach* 10686 (K; M; SRGH).

Also in S. Africa. In disturbed vegetation, on roadsides, along river banks, usually in sandy soil.

3. PTERODISCUS Hook.

Pterodiscus Hook. in Curtis, Bot. Mag. 70, t. 4117 (1844).
Pedaliophyton Engl. in Bot. Jahrb. 32: 111, t. 5 (1902).—Bruce in Kew Bull. 8: 419 (1953).

Small herbs, rarely more than 30 cm. high, with a persistent basal organ, consisting either of a short swollen aerial stem arising from a subterranean tuber of approximately the same diam., or a woody, not distinctly swollen stem (rarely

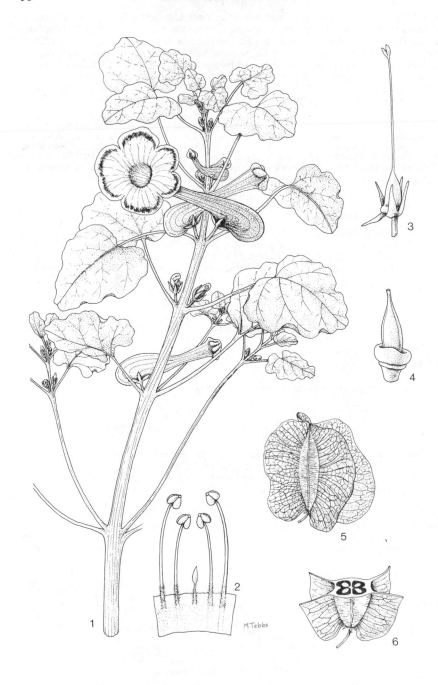

Tab. 20. HOLUBIA SACCATA. 1, flowering branch ($\times\frac{1}{2}$); 2, stamens, the posterior one represented by a staminode ($\times$1); 3, calyx and pistil ($\times$1); 4, ovary and disk ($\times$2), 1−4 after Hook., Ic. Pl. 15, tab. 1475 (1884); 5, fruit ($\times\frac{1}{2}$); 6, transverse section ($\times\frac{1}{2}$), 5−6 from *Hartmann* 1200.

two or three stems) arising from a napiform or pyriform subterranean tuber; several usually unbranched annual shoots produced from the top of the basal organ annually. Leaves subsucculent, very variable in shape, linear to broadly oblong, entire, undulate, dentate, pinnatifid or pinnatipartite. Flowers solitary in the leaf axils, yellow, brilliant orange, red or purple. Calyx small, persistent. Corolla tube funnel-shaped or narrowly cylindrical and constricted in the lower part, often slightly gibbous at the base (reduced spur); limb spreading; lobes subequal or distinctly different in size, circular to oblate. Stamens 4, included in the tube; thecae divergent. Ovary bilocular; loculi undivided; ovules 1−3, or (not in the Flora Zambesiaca area) 5−6 in each loculus, pendulous. Fruit indehiscent, laterally slightly compressed, with a 1−20 mm. broad longitudinal parchment-like wing on each of the four edges (sometimes rather inconspicuous); mesocarp spongy with large cavities, the thin endocarp sclerified and very tough; fruit oblate, circular, rotund or ovate in lateral view, often asymmetric, emarginate or apiculate at the apex, usually emarginate at the base; upper part of the fruit sterile and forming a distinct beak. Seeds variable in both shape and structure of the testa.

A genus of approximately 10 species, all native of Africa.

Pterodiscus species are very variable, especially in the fruit characters; fruits formed at the beginning or end of the season may be less than half the size of fruits formed in the middle of the season, and even their shape may be different; figures given in the descriptions below refer to fully developed fruits from the middle of the season. None of the species of the Flora Zambesiaca area occurs sympatrically with any other species of the genus, but where the distribution areas are in contact, numerous transitional forms, obviously due to introgressions, occur; correct determination is therefore difficult, unless the material is complete, i.e. comprises the basal organ, leaves, flowers and fully developed fruits.

1. Fruit ovate in lateral view, 10−15 mm. long, 6−8 mm. broad at the base, only very slightly compressed; the wings very narrow (0.5−1 mm. broad); a separate beak not discernable; the faces of the body often tuberculate; leaves narrowly oblong-lanceolate, up to 6 cm. long, 0.6−1.2 cm. broad, slightly undulate to slightly dentate; flowers dark yellow; corolla tube nearly cylindrical, 25−30 mm. long, limb c. 20 mm. in diam., the lobes subequal; the basal organ consisting of 1−3 very short aerial stems arising from a napiform subterranean tuber - - - - - - 1. *angustifolius*
− Fruit rotund, circular or oblate in lateral view, usually more than 15 × 8 mm., the wings conspicuous, at least 2 mm. broad - - - - - - - 2
2. Basal organ a swollen stem arising from a subterranean tuber of approximately the same diam.; corolla tube cylindrical; corolla either yellow with yellow, red or purplish subequal lobes, or brilliant orange with dark red throat and enlarged anterior lobe - - - - - - - - - - - - - - - 3
− Basal organ a short woody stem, not distinctly swollen, arising from a subterranean usually pyriform tuber; corolla wine-red to purple, sometimes with yellow throat; tube funnel-shaped or cylindrical - - - - - - - - 4
3. Corolla yellow, with yellow, red or purplish lobes, tube sometimes suffused with purple; the diam. of the limb not exceeding twice the diam. of the throat; lobes subequal, oblate; fruit usually rotund to circular in lateral view; beak distinct, 2−3 mm. broad; the wings of the two sides of the fruit not contiguous at the base of the fruit, the base therefore distinctly cordate - - - - - 2. *ngamicus*
− Corolla brilliant orange, with dark red throat; the diam. of the limb at least three times the diam. of the throat; anterior lobe of limb enlarged, nearly circular; fruit oblate in lateral view; beak very narrow, almost needle-like; the wings of the two sides of the fruit usually contiguous at the base of the fruit or often even overlapping; the ripe fruit often suffused with purple - - - - - 3. *aurantiacus*
4. Corolla tube funnel-shaped; lobes subequal; limb 35−40 mm. in diam., length of tube 25−30 mm.; leaves lanceolate to oblong, slightly undulate to slightly dentate; if dentate, then the teeth more conspicuous and concentrated towards the apex of the leaf; fruit rotund in lateral view, with distinct and very narrow beak 4. *elliottii*
− Corolla tube cylindrical; anterior lobe enlarged; limb 25−50 mm. in diam., length of tube 30−45 mm.; leaves usually pinnatilobed to pinnatifid; fruit ovate to rotund in lateral view, a separate beak not discernable - - - - 5. *speciosus*

1. **Pterodiscus angustifolius** Engl. in Bot. Jahrb. **19**: 155 (1894).—Engler, Pflanzenw. Ost-Afr.: 364 (1895).—Stapf in F.T.A. 4, 2: 545 (1906).—Bruce in Kew Bull. **1953**: 419 (1953).—Bruce in F.T.E.A., Pedaliaceae: 7 (1953). Type from Tanzania.
  Pedaliophyton busseanum Engl. in Bot. Jahrb. **32**: 111, t. 5 (1902).—Bruce in Kew Bull. **8**: 419 (1953). Type from Tanzania.

Pedalium busseanum (Engl.) Stapf in F.T.A. 4, 2: 540 (1906).—Bruce in Kew Bull. 1953: 419 (1953). Type as above.

Perennial herb, 10−20 cm. high, the basal organ consisting of 1−3 very short woody aerial stems arising from a napiform subterranean tuber. Leaves narrowly oblong to lanceolate, up to 6 cm. long, 0.8−1.2 cm. broad, slightly undulate to slightly dentate. Flowers dark yellow; corolla tube nearly cylindrical, 25 −30 mm. long, limb c. 20 mm. in diam., the lobes subequal, oblate. Fruit ovate in lateral view 10−15 mm. long, 6−8 mm. broad at the base, only very slightly compressed; wings 0.5−1 mm. broad, the faces of the body of the fruit often tuberculate. Seeds 1−2 in each loculus.

Mozambique. N: Erati, 6.5 km. from Namapa to Odinepa, 320 m., fl. & fr. 12.xii.1963, *Torre & Paiva* 9530 (LISC).
Also in Tanzania. In black soil in rock crevices.

2. **Pterodiscus ngamicus** N.E. Br. ex Stapf in F.T.A. 4, 2: 543 (1906).—Stapf in F.C. 4, 2: 456 (1904) (*P. "luridus"*).—Ihlenf. in Mitt. Bot. Staatss. München 6: 602−605 (1967) (*P. "luridus"*).—Merxm. & Schreiber in Merxm., Prodr. Fl. SW. Afr. 131: 5−6 (1968) (*P. "luridus"*). TAB. 21 fig. A1; C. Type: Botswana, Lake Ngami, fr. s. dat. *Chapman* s.n. (K, lectotype).

Perennial herb, up to 30 cm. high; the basal organ a swollen stem, 2−3 cm. in diam., sometimes divided at the top, arising from a subterranean tuber of approximately the same diam. Leaves very variable, up to 10 cm. long and 1.2−4.0 cm. broad, subentire, dentate or pinnatilobed, usually with 3 pairs of lateral veins. Corolla tube yellow, sometimes suffused with purple; tube cylindrical, 2.5−3.5 cm. long, limb c. 2.0 cm. in diam., not exceeding twice the diam. of the throat; lobes yellow, red or purplish, subequal, oblate; throat yellow. Fruit usually rotund to circular in lateral view, 2.5−3.0 cm. long, 2.5−2.8 cm. broad; beak distinct, 2−3 mm. broad; the wings of the two sides of the fruit not contiguous at the base of the fruit, base of the fruit therefore distinctly cordate. Seeds usually one in each loculus.

The original description of the flower is wrong since it was obviously taken from a specimen that belongs to *Harpagophytum*. This widespread species has been misinterpreted (since Stapf F.C. 4, 2: 456 (1904)) as *Pterodiscus luridus* Hook. f.; the true *P. luridus* has a very restricted distribution area near the southern coast of S. Africa. Unfortunately the type specimen of *P. ngamicus* possesses fruits that are not typical of this species, but are in shape and structure very close to those of *P. aurantiacus*; this is obviously due to the fact that the type locality is near the borderline between the distribution areas of *P. ngamicus* and *P. aurantiacus*.

Botswana. N: Okavango Delta, Toteng, fr. 20.i.1975, *Astle* 7134 (SRGH). SW: Ghanzi, fl. & fr. 19.x.1969, *Brown* 7008 (K; SRGH). SE: 8 km. S. of Palapye Rd., fr. 20.i.1960, *Leach & Noel* 266 (K; SRGH). **Zimbabwe.** W: Hwange Game Reserve, fr. 18.ii.1956, *Wild* 4745 (K; SRGH). E: Rupisi Hot Springs fl. & fr. 4.ii.1950, *Chase* 3546 (BM; SRGH). S: Near Shashi R., fl. 7.i.1898, *Rand* 197 (BM). **Mozambique.** GI: Mabalane, fl. & fr. 1.x.1963, *Leach & Bayliss* 11781 (K; SRGH). M: Goba, fl. & fr. 6.ii.1979, *de Koning* 7356 (WAG).
Also in S. Africa and Namibia. In sandy soil, often calcareous soil, and in clay, locally frequent.

3. **Pterodiscus aurantiacus** Welw. in Trans. Linn. Soc. 27: 53 (1869).—Hiern, Cat. Afr. Pl. Welw. 1: 795−796 (1900).—Stapf in F.T.A. 4, 2: 543 (1906).—Ihlenf. in Mitt. Bot. Staatss. München 4: 602−605.—Merxm. & Schreiber in Merxm., Prodr. Fl. SW. Afr. 131: 5−6. Type from Angola.
Pterodiscus brasiliensis sensu Schinz, Deutsch-Südwest-Afrika: 264 (1891) (*P. "brasiliensis"* (Gay) Aschers.).

Perennial herb, up to 30 cm. high, the branches erect or ascending; the basal organ a swollen stem arising from a subterranean tuber of approximately the same diam. Leaves lanceolate, up to 8 cm. long and 2 cm. broad, subentire, undulate or slightly dentate, usually with 4 pairs of lateral veins. Flowers brilliant orange red, with dark red throat; corolla tube cylindrical, 16−20 mm. long; diam. of limb at least three times the diam. of the throat; anterior lobe of limb enlarged, nearly circular. Fruit usually oblate in lateral view, up to 35

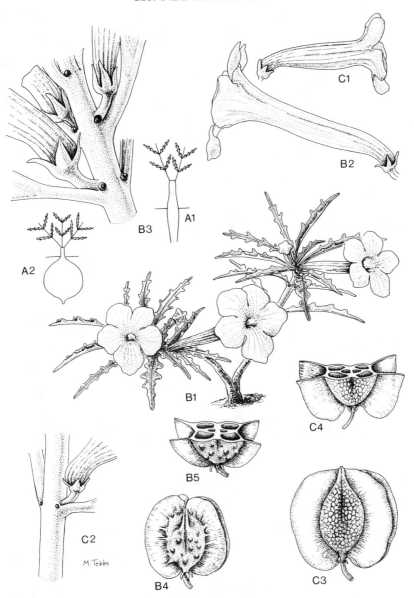

Tab. 21. A.—PTERODISCUS. Two types of persistent basal organs. A1, swollen aerial stem arising from a subterranean tuber of approx. the same diam.; A2, woody, not distinctly swollen stem arising from a napiform or pyriform subterranean tuber; unbranched annual shoots are produced at the apex of the basal organs. B.—PTERODISCUS SPECIOSUS. B1, habit (×½), based on a life photograph (*Ihlenfeldt* 2090); B2, flower, lateral view (×1); B3, part of annual shoot showing the lower portions of the flowers in the leaf axils and the large dark green extrafloral nectaries with droplets of nectar; B2—3 based on photographs (*Ihlenfeldt* 2027); B4, fruit, lateral view (×1); B5, fruit, transverse section (×1), B4—5 from *Ihlenfeldt* 2090. C.—PTERODISCUS NGAMICUS. C1, flower, lateral view (×1); C2, part of an annual shoot showing the lower portion of flower and green extrafloral nectary with a droplet of nectar (×2), C1—2 based on photographs (*Ihlenfeldt* 2120); C3, fruit, lateral view (×1); C4, transverse section (×1), C3—4 from *Ihlenfeldt* 2120.

mm. long and 41 mm. broad, often suffused with purple when ripe; beak very narrow, almost needle-like; the wings of either side of the fruit contiguous or even overlapping at the base of the fruit. Seeds 1—2 in each loculus.

Botswana. SW: Gate between Namibia and Botswana, fl. 13.ii.1970, *Brown* in *Mamuno* 38 (K; SRGH).
Also in Angola and Namibia. In clayey soil under shrubs.

4. **Pterodiscus elliottii** Baker ex Stapf in F.T.A. 4, 2: 542 (1906). TAB 21 fig. A2. Type: Zimbabwe, Matabeleland without precise locality, fl. ii.1886, *Elliott* s.n. (K).

Perennial herb, c. 20 cm. high; the basal organ consisting of a short woody stem arising from a subterranean pyriform or sometimes napiform tuber. Leaves lanceolate or oblanceolate, up to 12 cm. long and 2.5 cm. broad, undulate to slightly dentate; if dentate, then the teeth more conspicuous and concentrated towards the apex of the leaf, usually with 4—5 pairs of lateral veins. Flowers dark purple to wine-red, sometimes with a yellow throat; corolla tube funnel-shaped (except at the very base), 40—50 mm. long, limb c. 35 mm. in diam.; lobes subequal. Fruit rotund in lateral view, c. 22 mm. long and 20 mm. broad; the wings c. 3—4 mm. broad; beak very narrow. Seeds 1—2 in each loculus.

The original description is based on a very poor specimen consisting of a single stem with flower, without the basal organ and without fruits which are described here for the first time.

Zambia. E: Chipata, Mkania, fl. 12.xii.1968, *Astle* 5387 (SRGH). N: Isoka, fl. 2.i.1963, *Fanshawe* 7456 (SRGH). **Zimbabwe.** N: Urungwe Distr, Mensa Pan, fr. 30.i.1958, *Drummond* 5360 (BR; K; LISC; SRGH). W: Khami Ruins near Bulawayo, fl. x.1957, *Leach* 8169 (K; SRGH). C: Harare, Agriculture Expt. Stat. fl. 30.xii.1941, *Pardy* s.n. (SRGH).
In woodland, especially with *Colophospermum*

5. **Pterodiscus speciosus** Hook. in Curtis, Bot. Mag. 70, t. 4117 (1844).—Stapf in F.C. 4, 2: 456 (1904); in F.T.A., 4, 2: 542 (1906). TAB 21 fig. A2, B. Type from S. Africa.
    Harpagophytum pinnatifidum Engl., Bot. Jahrb. 10: 255 (1889).—Ihlenf. in Mitt. Bot. Staatss. München 6: 605 (1967). Type from S. Africa.

Perennial herb, up to 20 cm. high; the basal organ a short woody stem arising from a subterranean usually pyriform tuber. Leaves lanceolate, up to 6 cm. long and 1.5 cm. broad, usually pinnatilobed and with 5 pairs of lateral veins. Flowers red-purple; tube cylindrical, 25—50 mm. long, limb 30—45 mm. in diam.; lobes with darker stripes leading to the dark red throat; anterior lobe enlarged. Fruit ovate to rotund in lateral view, normally c. 18 mm. long and c. 16 mm. broad, but sometimes up to 22 mm. long; the wings c. 3 mm. broad; a separate beak not discernable. Seeds one in each loculus.

Botswana. SE: Gaberones, near Content Farm, fl. & fr. 31.i.1973, *Kelaole* A128 (SRGH).
Also in S. Africa. In alluvial soil in grassland.

Most of the gatherings from Botswana SE. are transitional forms between *P. speciosus* and *P. ngamicus*.

4. CERATOTHECA Endl.

Ceratotheca Endl. in Linnaea 7: 5, t. 1—2 (1832).—Abels in Mém. Soc. Brot. 25: 1—358 (1975).
Sporledera Bernh. in Linnaea 16: 41—42 (1842) non. Hampe (1837).—Benth. in Gen. Pl. 2, 2: 1059 (1876).

Annual or short-lived perennial herbs, or (not in the Flora Zambesiaca area) small shrubs, erect, branched, sometimes semi-prostrate; leaves and stems more or less densely covered with mucilage-glands and hairs. Leaves opposite, petiolate; lamina very variable, circular, ovate, cordate, reniform or lanceolate, hastate or tri-lobed, often polymorphic, margins nearly entire, crenate or serrate; (superior and inferior leaves often much smaller and simplified). Flowers shortly pedicellate, solitary in the leaf-axils, pendent, white, pink or lilac, the throat usually less intensively coloured. Corolla tube funnel-shaped, the base on the adaxial side slightly gibbous; limb sub-bilabiate; the upper and lateral lobes short and subequal, the anterior lobe enlarged, often yellow, with dark stripes or lines of dark spots running down into the throat. Stamens four, included, inserted near the base of the tube; staminodium absent; thecae parallel. Ovary bilocular, each loculus divided by a false septum almost to the apex, each compartment with numerous ovules, uniseriate. Fruit a loculicidally dehiscent capsule, laterally compressed, obtuse or truncate at the apex, with two lateral horns (sometimes not very conspicuous) at the angles of the apex. Seeds numerous, spreading or ascending, obovate in outline, usually black when ripe.

A genus of 5 species, all African.

Leaves (in the middle portion of the stem) lanceolate-deltate or ovate-cordate, usually coarsely dentate at least towards the base, 1.5−8 cm. long, 0.4−4.5 cm. broad, acute at the apex, truncate, broadly cuneate or subhastate at the base; corolla 1.3−4 cm. long; capsule 0.8−2.1 cm. long, 0.4−0.7 cm. broad, with usually horizontal or slightly ascending slender (sometimes inconspicuous) horns; seeds ascending, broadly obovate in outline, with a single fringe; testa in the central portion smooth, in the marginal portion radially rugose - - - - - - - - - - 1. *sesamoides*
Leaves (in the middle portion of the stem) broadly cordate or trilobed, sometimes even tripartite or digitate, occasionally 5−7-lobed, the margin coarsely crenate, the lamina up to 15 cm. broad and long; corolla 2.0−6.1 cm. long; capsule 1.5−2.8 cm. long and 0.5−0.7 cm. broad; horns c. 8 mm. long, each with broad triangular base, horizontal; seeds spreading, obovate in outline, with a double fringe, testa (except for the fringe) reticulate - - - - - - - - - - 2. *triloba*

1. **Ceratotheca sesamoides** Endl. in Linnaea 7: 5−8, t. 1−3 (1832).—Stapf in F.T.A. 4, 2: 563 (1906).—Bruce in F.T.E.A., Pedaliaceae: 14, fig. 6. (1953).—Heine in F.W.T.A., ed. 2, 2: 391 (1963).—Ihlenf. in Mitt. Bot. Staatss. München 6: 594−598 (1967).—Merxm. & Schreiber in Merxm., Prodr. Fl. SW. Afr. 131: 2−3 (1968).—Abels in Mem. Soc. Brot. 25: 1−358 (1975).—Hakki in Fl. Anal. Togo: 383 (1984). TAB. 22 fig. B. Type from Senegal.
Ceratotheca melanosperma Hochst. ex Bernh. in Linnaea 16: 41 (1842).—Stapf in F.T.A. 4, 2: 563 (1906). Type from Sudan.
Ceratotheca sesamoides var. *melanoptera* A.DC. in DC. Prodr. 9: 252 (1845).—Stapf in F.T.A. 4, 2: 563 (1906). Type as above.
Ceratotheca sesamoides f. *latifolia* Engl., Bot. Jahrb. 32: 115 (1903).—Abels in Mem. Soc. Brot. 25: 219 (1975). Lectotype from Togo.
Sesamum heudelotii Stapf, F.T.A. 4, 2: 552 (1906).—Heine in F.T.W.A., ed. 2, 2: 391 (1963). Type from Senegambia.

Annual herb, simple or branched, erect or semi-prostrate, up to 1.2 m. high, pubescent. Leaves (in the middle portion of the stem) lanceolate-deltate or ovate-cordate, usually coarsely dentate at least towards the base, 1.5−8 cm. long, 0.4−4.5 cm. broad, acute at the apex, truncate, broadly cuneate or subhastate at the base, subsessile or with petioles up to 6 cm. Flowers pink, lilac, mauve or occasionally purple, the throat and lip often cream with dark lines; corolla 1.3−4 cm. long, thinly pubescent; anterior lobe broadly ovate, the upper and lateral lobes much shorter and transversely elliptic. Capsule pubescent and glandular, 0.8−2.1 cm. long, usually laterally much compressed; horns slender (sometimes inconspicuous), 1−3.3 mm. long, usually horizontal or slightly ascending. Seeds ascending, broadly obovate in outline, with a single fringe, c. 2.7 mm. long and 2.2 mm. broad, the central portion convex; testa in the central portion smooth, in the marginal portion radially rugose.

Caprivi Strip: Mpola, 24 km. from Katima Mulilo on Rd. to Ngoma, fl. 5.i.1959, *Killick & Leistner* 3294 (K; SRGH; WIND). **Botswana.** N: Ngamiland, Pandamatenga, fl. & fr. 28.iii.1961, *Richards* 14895 (K; SRGH). **Zambia.** B: Mongu Distr., fl. 10.i.1960, *Gilges* 953 (K; SRGH). N: Lake Tanganyika, Mpulungu, fl. & fr. 6.iii.1952, *Richards* (K). W: 0.8 km. from Kafue R., Kitwe, fl. 19.iii.1961, *Linley* 122 (SRGH). C: Iolanda Farm, near Kafue town, fl. & fr. 14.iii.1965, *Robinson* 6437 (M). E: Chipata, fl. & fr. 3.vi.1958, *Fanshawe* F4516 (K). S: Mazabuka, c. 1.5 km. from Zambezi, fl. & fr. 23.xii.1956, *Scudder* 7 (SRGH). **Zimbabwe.** N: N. side of R. Mavora, eastern Urungwe, fl. & fr. 26.ii.1958, *Phipps* 888 (BR; SRGH). W: Hwange Distr., fl. & fr. vi.1941, *Sturgeon* s.n. (SRGH). E: Between Musavasivi R. and Cilariati R., fl. & fr. 22.i.1957, *Phipps* 119 (SRGH). S: Masvingo Distr., Triangle Ranch, fl. 15.i.1947, *Bates* s.n. (SRGH). **Malawi.** N: Khondwe to Karonga, fl. & fr. vii.1896, *Whyte* s.n. (K). C: Ntchisi Distr., Chitembwe, fl. 4.iii.1939, *Herklots* 54 (EA). S: Mangochi, fl. & fr. 21.iii.1956, *Jackson* 1832 (K). **Mozambique.** N: Nampula, fl.

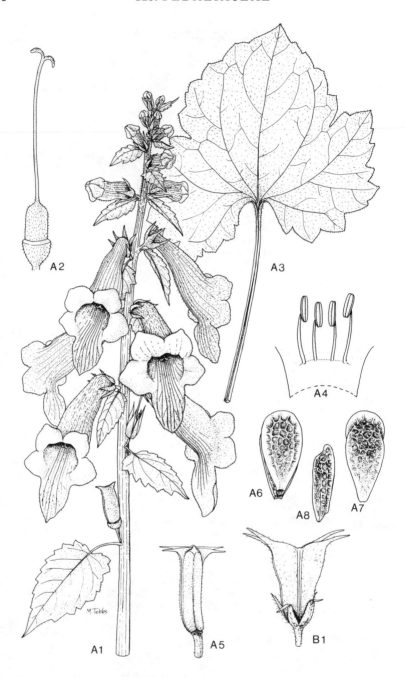

Tab. 22. A.—CERATOTHECA TRILOBA. A1, flowering branch (×½); A2, ovary and pistil (×1); A3, leaf from middle portion of stem (×½); A4, opened flower, showing insertion of stamens (×1), A1—4 after Curtis, Bot. Mag. Ser. 3, **44**: t. 6972 (1888); A5, fruit, lateral view (×1); A6, seed, superior surface (×8); A7, seed, inferior surface (×8); A8, seed, lateral view (×8), A5—8 from *Ihlenfeldt* 2118. B.—CERATOTHECA SESAMOIDES. B1, fruit, lateral view (×½), from F.T.E.A.

& fr. 6.ii.1937, *Torre* 1265 (LISC). Z: Between Serra de Murrumbala and Chire R., fl. & fr. 13.xii.1971, *Pope & Müller* 603 (SRGH). T: Changara, c. 8 km. along Tete Rd., fl. & fr. 11.vii.1958, *Seagrief* 3052 (LISC; SRGH). MS: Ancueza, C.I.C.A. Experimental Station fl. & fr. 14.iv.1960, *Lemos & Macuacua* 87 (K; LISC; PRE; SRGH). GI: Govuro, fl. & fr. 18.iii.1974, *Correia & Marques* 4041 (WAG).

Also in N. Africa from Senegal to Tanganyika, Zaire, Namibia. In open grassland and tree savanna on sandy soils, rarely among rocks, often in disturbed vegetation, sometimes recorded as a weed. It is planted like *Sesamum* for the seeds; the leaves are used as a vegetable.

2. **Ceratotheca triloba** (Bernh.) Hook. f. in Curtis, Bot. Mag. Ser. 3, **44**: t. 6974 (1888). TAB. 22 fig. A. Type from S. Africa.

Sporledera kraussiana Bernh. in Linnaea **16**: 41 (1842).—Benth. in Gen. Pl. 2, 2: 1059 (1876). Type from S. Africa.

Sporledera triloba Bernh. in Linnaea **16**: 42 (1842).—Stapf in F.C.4, 2: 462 (1904).—Stapf in F.T.A. 4, 2: 564 (1906).—Abels in Mem. Soc. Brot. 25: 1−358 (1975). Type from S. Africa.

Sesamum lamiifolium Engl., Bot. Jahrb. **10**: 256, t. 8 (1898).—Stapf in F.C. 4, 2: 462 (1904). Type as above.

Ceratotheca lamiifolia (Engl.) Engl., Bot. Jahrb. **19**: 156 (1894). Type from S. Africa.

Annual or perennial herb, up to 2.5 m. high, erect and branched; stems pubescent; whole plant malodorous. Leaves (in the middle portion of the stem) broadly cordate or tri-lobate sometimes even tripartite or digitate, occasionally 5−7 lobed, margin coarsely crenate, pubescent; lamina up to 15 cm. broad and long; petiole up to 13 cm. long. Flowers lilac to cream with purple stripes on the anterior lobe running down into the throat. Corolla 2.0−6.1 cm. long, the anterior lobe 5−7 mm. long, elliptic-ovate; the lateral and upper lobes subequal, broadly ovate. Capsule 1.5−2.8 cm. long and 0.5−0.7 cm. broad, laterally only slightly compressed; horns c. 8 mm. long, each with broad triangular base, usually horizontal. Seeds spreading, obovate in outline, with a double fringe, c. 2.8 mm. long and 1.7 mm. broad, the central portion plane; testa (except for the fringe) reticulate.

Botswana. SW: On Rd. from Khakhea to Kanye, fl. & fr. 24.ii.1960, *de Winter* 7504 (M; PRE; SRGH). SE: c. 120 km. W.N.W. of Francistown, fl. & fr. 2.v.1957, *Drummond* 5291 (K; SRGH). **Zimbabwe**. W: Bulawayo, fl. & fr. v.1915, *Rogers* 13658 (G; K; Z). C: Marondera, fl. 12.iii.1942, *Dehn* 187 (M; SRGH). E: Mutare, fl. ii.1959, *Head* s.n. (BM). S: Lundi R., fl. & fr. 30.vi.1930, *Hutchinson & Gillett* 3268a (K; LISC; SRGH). **Mozambique**. GI: Near Mapulanguene, fl. & fr. 17.ii.1953, *Myre & Balsinhas* 1527 (K; LISC; SRGH). M: Namaacha Gorge, fl. & fr. 4.iv.1941, *Hornby* 700 (EA).

Also in S. Africa. Cultivated as an ornamental plant in warm countries. In sandy soils, often in disturbed vegetation.

5. SESAMUM L.

Sesamum L., Sp. Pl. ed. **1**: 634 (1753); Gen. Pl., ed. 5: 282 (1754).

Simsimum Bernh. in Linnaea **16**: 38 (1842).

Ganglia Bernh. in Linnaea **16**: 39 (1842).

Sesamopteris DC., Prodr. **9**: 251 (1845).

Volkameria L. ex Kuntze, Rev. Gen. Pl. **2**: 481 (1891).

Annual or perennial, erect herbs, sometimes shrubs (not in the Flora Zambesiaca area) Leaves sessile or petiolate, lamina entire, lobed or digitate, often varying on the same plant. Flowers solitary in the leaf axils, shortly pedicellate with extra-floral nectaries (reduced flowers) at the base. Calyx persistent or deciduous. Corolla white, pink or purple, obliquely campanulate, limb sub-bilabiate, lowest lobe the longest. Anthers dorsifixed, thecae parallel, connective gland-tipped. Disk annular, regular. Ovary bilocular, subcylindrical; loculi divided almost to the apex by a false septum; ovules numerous, uniseriate in each division. Fruits longitudinally dehiscent capsules, cultrate to narrowly oblong or obconical in lateral view, 4-sulcate, rostrate at the apex. Seeds numerous, obovate, compressed, winged or with a double, rarely single fringe, testa smooth or rugose.

A genus subdivided into 4 sections (Ihlenf. & Seidenst. in Proc. IX Plen. Meet. A.E.T.F.A.T. Las Palmas: 53−60 (1979)), comprising 21 taxa of specific rank; the majority of species are distributed through tropical and subtropical Africa, one section (*Chamaesesamum* Benth.) is restricted to India and Sri Lanka.

Key to the Sections

1. Seeds winged, testa faveolate; bracts of extra-floral nectaries minute; inferior lip of the flower not or little prominent (p. 000) - - - - - 2. **Sesamopteris**
   — Seeds not winged but with a double fringe; bracts of extra-floral nectaries conspicuous; inferior lip of the flower often prominent - - - - - - - 2
2. All leaves entire; inferior lip of the flower prominent; testa sculptured in various ways (p. 000) - - - - - - - - - - - 1. **Aptera**
   — Lower leaves lobed, digitate or entire; inferior lip of the flower not prominent; testa smooth or faintly reticulate (p. 000) - - - - - - - 3. **Sesamum**

Sect. **Aptera** Seidenst. in op. cit.

Erect perennial or annual herbs, leaves entire. Flower with a prominent inferior lip; extra-floral nectaries large, single, more or less stalked, with two conspicuous bracts. Pollen grains with 7−9 colpi. Capsules usually neither ribbed nor with protruding loculi. Seeds not winged, but with a double fringe; testa sculptured in various ways, but never faveolate, brown to black when ripe.

1. Inferior surface of leaves tomentose with white hairs, obscuring the glands; corolla large, 3.5−7 cm. long; herb 0.8−3 m. high, stems more or less pubescent, leaves narrowly elliptic or oblanceolate, 2−11 × 0.4−4 cm.; corolla purple or pink; beak of the capsule broad and short (up to 2 mm. long) - - - - 1. *angolense*
   — Inferior surface of leaves usually not white tomentose; if white tomentose, beak of the capsule up to 6 mm. long; corolla smaller - - - - - - - 2
2. Seeds with a complete double fringe; testa rugose (TAB. 24 figs. B, C); herbs 0.5−2.0 m. high; leaves linear to linear-lanceolate, 2−12 × 0.2−1.8 cm. - - - 3
   — Seeds only in the lower part with a double fringe, in the upper part with a single fringe; testa on the superior face with radial stripes (TAB. 25 fig. D); the whole plant thinly pubescent; leaves glaucous; capsules narrowly oblong in lateral view, c. 2.5 × 0.8 cm. - - - - - - - - - - - - 4. *radiatum*
3. Capsules narrowly oblong in lateral view (TAB. 23 figs. B1−3), c. 2 × 0.4−0.6 cm., pubescent; seeds c. 2 × 1.3 mm. (TAB. 24 fig. B)) - - - - 2. *calycinum*
   — Capsules cultrate in lateral view, c. 2 × 0.3 cm., glabrescent; seeds with a very narrow double fringe (TAB. 24 fig. C), c. 1.3 × 1.0 mm., testa with transversal ribs (fig. xx) - - - - - - - - - - - - - 3. *angustifolium*

1. **Sesamum angolense** Welw., Apont.Phyto-Geogr.: 588 (1859); in Trans. Linn. Soc. 27: 51 (1869).—Stapf in F.T.A. 4, 2: 555 (1906).—Bruce in F.T.E.A., Pedaliaceae: 20 (1953).—Ihlenf. & Seidenst. in Mitt. Bot. Staatss. München 7: 11−12 (1968). TAB. 23 fig. A, TAB 24 fig. A. Type from Angola.
   Sesamum macranthum Oliver in Trans. Linn. Soc. 29: 131 (1875).—Bruce in F.T.E.A., Pedaliaceae: 21 (1953). Type from Tanzania.
   Sesamum macranthum var.? *angustifolium* Oliver, loc. cit.—Bruce loc. cit. Type from Tanzania.

Erect annual or perennial herb, simple or branched, 0.8−3 m. high, malodorous. Leaves subsessile or shortly petiolate; lamina cultrate, narrowly oblong, narrowly elliptic or oblanceolate, 2−11 × 0.4−4 cm., margins entire, more or less involute cuneate at the base, truncate, retuse or rarely acute, and usually mucronate at the apex, inferior surface white tomentose. Flowers reddish, pink, or pale mauve with darker markings; calyx lobes narrowly lanceolate or narrowly ovate, 5−10 mm. long, about 2 mm. broad at the base, acute or acuminate at the apex, pubescent; corolla large, 3.5−7 cm. long, 2−3 cm. in diam. at the throat. Capsule narrowly oblong in lateral view, subquadrangular, 4-sulcate, 18−28 × 5−7 mm., rather densely pubescent, gradually narrowed into a flattened rather broad and short beak (TAB. 23 fig. A). Seeds with a double fringe, about 2 × 1.5 mm., faintly rugose on the flanks and faces (TAB. 24 figs. A1, A2).

Zambia. N: By stream c. 11.27 km. N.W. of Mbala, fl. & fr. 19.vii.1930, *Hutchinson & Gillett* 3920 (BM; LISC; SRGH). W: Ndola, Kasongo, fl. 14.vii.1953, *Fanshawe* 146 (LISC). C: 13 km. N. of Mchinji, fl. & fr. 30.v.1961, *Leach & Rutherford-Smith* 11078 (SRGH). S: Mazabuka, fl. & fr. 28.v.1963, *van Rensburg* 2231 (K). **Zimbabwe.** N: Urungwe Distr., Kapondo Urengi R., fl. & fr. 14.v.1954, *Lovemore* 396 (K). **Malawi.** N: Vicinity of Uzumara Forest, fr. v.1953, *Chapman* 155 (BM). C: Lilongwe Distr., Nsaru, fl. 22.iii.1955, *Exell, Mendonça & Wild* 1130 (BM; LISC; SRGH). S: Zomba Township, fl. 29.ii.1956, *Banda* 196 (BM). **Mozambique.** N: Along Rd. from Mahua to Marrupa, c. 68 km. from Mahua, fr. 14.viii.1981, *Jansen, de Koning & de Wilde* 300 (BM).
   Also in Angola, Kenya, Uganda, Tanzania and Zaire. Common in grassland and open woodland, by roadsides, and in abandoned cultivation.

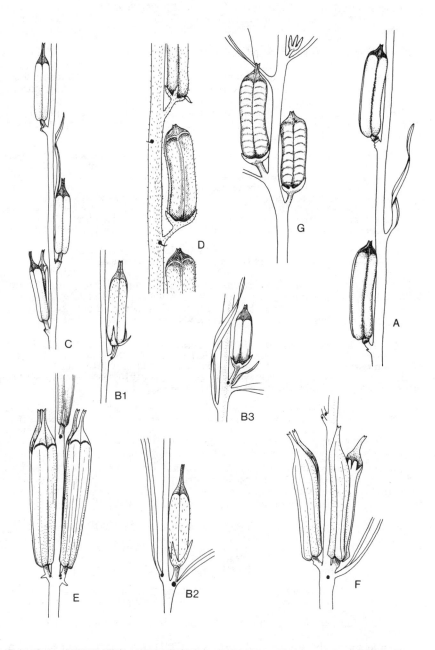

Tab. 23. A.—SESAMUM ANGOLENSE. A, Fruits, lateral view (×1), from *Chapman* 155.
B.—SESAMUM CALYCINUM Subsp. CALYCINUM. B1, fruit (×1) from *Read* 29;
B2, fruit of SESAMUM CALYCINUM Subps. PSEUDOANGOLENSE (×1), from
Jack Herb. No. 43777.B3 fruit of Subsp. BAUMII (×1), from *Gilges* 710A.
C.—SESAMUM ANGUSTIFOLIUM. C, fruit (×1), from *Tyrer* 272. D.—SESAMUM
RADIATUM. D, fruits (×⅓), from *Santo* 899. E.—SESAMUM ALATUM. E, fruits
(×⅗), from *Codd* 7347. F, fruits of SESAMUM TRIPHYLLUM (×1), from *Hopkins*
Herb No. 7781. G.—SESAMUM INDICUM. G, fruits (×1 1/10), from *Swynnerton*
251.

2. **Sesamum calycinum** Welw. in Trans. Linn. Soc. **27**: 52 (1869).—Stapf in F.T.A. 4, 2: 555 (1906).—Merxm. & Schreiber in Merxm., Prodr. Fl. SW. Afr. **131**: 11 (1968).— Ihlenf. & Seidenst. in Mitt. Bot. Staatss. München **7**: 11—12 (1968). Type from Angola.

Annual or perennial, erect or sometimes ascending herb; stem simple or branched. Leaves linear to cultrate, narrowly lanceolate to lanceolate, entire, inferior surface glabrous - except for the mucilage glands - or rarely pilose (subsp. *pseudoangolense*). Flowers obliquely campanulate, corolla 2—5 cm. long, purple, mauve, pink or white; calyx persistent. Capsule narrowly oblong in lateral view, c. 10—25 × 4—6 mm., 4-sulcate, glabrous or pilose, rostrate at the apex (TAB. 23 figs. B1—3). Seeds compressed, with a broad double fringe, black or brown when mature (TAB. 24 fig. B1—4).

1. Flowers deep purple, purple or mauve to violet, rarely pink; capsule pubescent, robust, 1.6—2.0 × 0.4—1.6 cm.; leaves not densely glandular on the inferior surface   2
 − Flowers pink to white; capsule glabrescent, parchment-like, 1.0—1.7 × 0.4—0.6 cm.; leaves densely glandular on the inferior surface -   -   -   - subsp. *baumii*
2. Corolla funnel-shaped, up to 4 cm. long, pink, mauve or purple; apical beak of the capsule obtuse (TAB. 23 fig. B1) -   -   -   -   -   - subsp. *calycinum*
 − Corolla campanulate, up to 5 cm. long, deep purple to violet; apical beak of the capsule slender and acuminate (TAB. 23 fig. B2 and Gilg) - subsp. *pseudoangolense*

Subsp. **calycinum** TAB. 23 fig. B1 & TAB. 24 fig. B1.
  Sesamum repens Engl. and Gilg loc. cit. Type from Angola.

Capsule narrowly oblong in lateral view, robust, matted pubescent, 16—20 × 3—4 mm., the apical beak straight, up 2—6 mm. long (TAB. 23 fig. B1) Seeds 1.5—2.4 × 1.1—1.6 mm.; testa faintly rugose, often with radial or transversal stripes or ribs (TAB. 24 fig. B1); seeds black or brown when mature.

Zambia. W: Bwana Mkubwa, fl. 7.ii.1954, *Fanshawe* 790 (BR). W: Kabwe, fl. & fr., s. dat., *Rogers* 9 (Z). S: Namwala Distr., fl. & fr. 1934/35, *Read* 29 (PRE). **Malawi**. C: Lilongwe Distr., Chitedze, fl. & fr. 22.v.55, *Exell, Mendonça & Wild* 1106 (SRGH).
Also in Angola and Zaire. On sandy soil, locality frequently by roadsides, in grassland and open woodland, often cultivated as a vegetable.

Subsp. **pseudoangolense** Seidenst* subsp. nov. TAB. 23 fig. B2 & TAB. 24 fig. B2. Type: Zimbabwe, Chegutu Distr., c. 70 km. ab urbe Harare versus pagum Kadoma, fl. & fr. 24.iii.1931, *Norlindh & Weimarck* 5141 (BM, holotype, BR, PRE, S, isotypes).
  Sesamum calycinum var. *calycinum* pro parte.—Ihlenf. & Seidenst. in Mitt. Bot. Staatss. München **7**: 12 (1968).

Capsule narrowly oblong in lateral view, robust, brownish, matted pubescent, 1.6—2.0 × 0.4—0.5 cm., the apical beak straight, 4—7 mm. long. Seeds black when mature, 1.8—2.2 × 1.3—1.6 mm.; testa with radial or tranversal ribs (TAB. 23 fig. B2). Corolla always deep purple, up to 5 cm. long.

Zimbabwe. N: Trelawney, Tabacco Exp. Station, fl. & fr. 23.xii.1942, *Jack* 71 (SRGH). W: Shangani, Gwampa Forest Reserve, fr. v.1956, *Goldsmith* 98/56 (K; PRE; SRGH). C: Marondera Distr., fl. & fr. 5.iv.1950, *Wild* 3252 (K; M; SRGH). E: Mutare Distr., Honde Valley, fl. & fr. ii.1964, *Chase* 1371 (BM; SRGH).
On sandy soil, frequent by roadsides, in grassland and at the edge of vleis.

Subsp. **baumii** (Stapf) Seidenst. stat. nov. TAB. 24 fig. B3, B4. Type: Angola, between Kiteve and Humbe, 2.vi.1900 *Baum* 959 (K, holotype. BM; BR; COI; E; M; S; W; Z, isotypes).
  Sesamum baumii Stapf in F.T.A. 4, 2: 554 (1906). Type from Angola.

Capsule narrowly oblong in lateral view, glabrous, parchment-like when mature, light coloured, 1.0—1.7 × 0.4—0.6 cm., the apical beak comparatively

*Capsula pubescens glabrescens, brunea, subgquadrangularis, quadrisulcate, apice in rostrum 4.0—5, 8—7.0 mm. longum contracta, rostro excluso 16.0—17.6—20.0 mm. longa, 4.0—4.4—4.8 mm. lata, basi in latere adaxiali attenuata. Semina nigrescentia, in regionibus marginalibus cum costis radialibus, in centro cum sculpturis irregularibus vel horizontalibus.
  A more extensive description will shortly be published in Mitt. Inst. Allg. Botanik, Hamburg, **22** (1988).

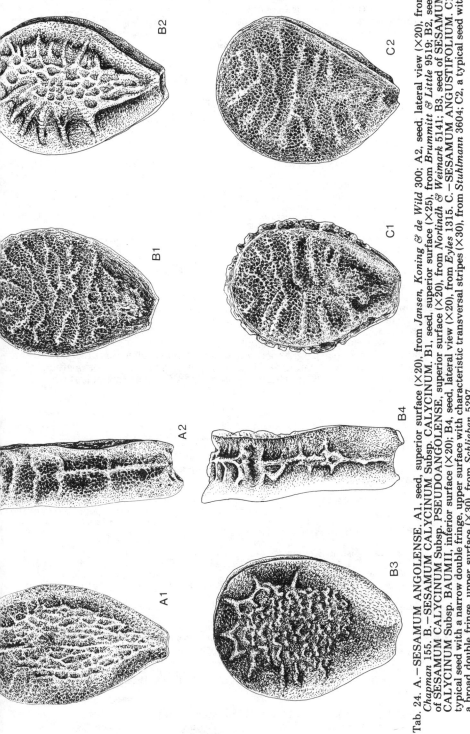

Tab. 24. A. – SESAMUM ANGOLENSE. A1, seed, superior surface (×20), from *Jansen, Koning & de Wild* 300; A2, seed, lateral view (×20), from *Chapman* 155. B. – SESAMUM CALYCINUM Subsp. CALYCINUM. B1, seed, superior surface (×25), from *Brummit & Little* 9519; B2, seed of SESAMUM CALYCINUM Subsp. PSEUDOANGOLENSE, superior surface (×20), from *Norlindh & Weimark* 5141; B3, seed of SESAMUM CALYCINUM Subsp. BAUMII, inferior surface (×20), from *Eyles* 1315. C. – SESAMUM ANGUSTIFOLIUM. C1, typical seed with a narrow double fringe, upper surface with characteristic transversal stripes (×30), from *Stuhlmann* 3604; C2, a typical seed with a broad double fringe, upper surface (×30), from *Schlieben* 5297.

short, 1.0−2.5−4.0 mm. long (TAB. 23 fig. B2). Seeds 1.7−2.4 × 1.2−1.6 mm., the upper fringe more or less bending upwards; testa of the upper face with radial or transversal stripes, often faintly rugose (TAB. 24); seeds black or brown when mature. Corolla always pink to white. Leaves greyish in appearance due to very densely packed mucilage glands.

Caprivi Strip: Lisikili, 24 km. E. of Katima Mulilo, fl. & fr. 17.vii.1952, *Codd* 7180 (PRE; BM). **Botswana**. N: without locality, fl. & fr. ? *Curson* 442 (PRE). **Zambia**. B: Kanda Pan, c. 12 km. NE. of Mongu, fl. & fr. 11.xi.1959, *Drummond & Cookson* 6339 (SRGH). S: Livingstone Distr., Kazungula, fl. & fr. 5.i.1957, *Gilges* 710a (SRGH). **Zimbabwe**. W: Victoria Falls, Long Island, fl. & fr. iv.1918, *Eyles* 1315 (SRGH).

Outside the Flora Zambesiaca area this subspecies is only found in localities along the Okavango R. in NE. Namibia and S. Angola. As a typical Zambezi-element it is common on sandy soil on riverbanks, also locally frequent in grassy places.

3. **Sesamum angustifolium** (Oliver) Engl., Pflanzenw. Ost-Afr. C: 365 (1895).—Stapf in F.T.A. 4, 2: 554 (1906).—Bruce in F.T.E.A., Pedalicaeae: 19 (1953). TAB. 23 fig. C & TAB. 24 fig. C. Type from Tanzania.
  Sesamum indicum var.? *angustifolium* Oliver in Trans. Linn. Soc. 29: 131 (1875). Type as above.
  Sesamum calycinum var. *angustifolium* (Oliver) Ihlenf. & Seidenst. in Mitt. Bot. Staatss. München 7: 12 (1968).Type as above.
  Sesamum baumii auct. non Stapf, Bruce in F.T.E.A., Pedaliaceae: 20 (1953). Type from Angola.

Erect annual or perennial herb, stem simple or branched, 0.4−2.0 m. high. Leaves sessile, subsessile or the inferior ones petiolate, lamina 4−12 × 0.1−2.0 cm., variable in shape, linear to cultrate, margins entire, more or less inrolled, or narrowly lanceolate to lanceolate, irregularly toothed, cuneate at the base, acute or obtuse at the apex, densely glandular on the inferior surface. Flowers purple, pink or mauve, often spotted within. Corolla 2.5−3.5 cm. long, 1−2 cm. in diam. at the throat. Capsule cultrate in lateral view, appressed to the stem (TAB. 23 fig. C), light coloured, parchment-like when mature, glabrescent, 15−25 × 3.5 mm.; beak of the fruit straight or bent towards the stem, up to 5 mm. long (TAB. 23 fig. C). Seeds on both faces with transversal ribs, c. 1.3 × 1mm., black when mature (TAB. 24 fig. C1).

Zambia. N: Kaputa Distr., near Kasongole Village, Katanga border, NW. of Mweru-wa-Ntipa swamp, fr. 3.viii.1962, *Tyrer* 272 (BM; SRGH). W: Kitwe, fl. & fr. 5.iv.1964, *Fanshawe* 8436 (SRGH). **Malawi**. N: Mzimba Distr. Mbawa Exp. Stat., fl. & fr. 5.iv.1955, *Jackson* 6 (SRGH). **Mozambique**. N: Maniamba, Mecaloge do Rovuma, fl. & fr. 3.ix.1934, *Torre* 537 (LISC).

Also in Kenya, Tanzania, Uganda and Zaire. Common by roadsides, in grassland and cultivated areas.

Typical *Sesamum angustifolium* occurs particularly in Kenya and Tanzania; the typical seed characters are often missing in plants from the bordering regions. Seeds of the representative populations of the Flora Zambesiaca area posses a broad double fringe rather than a very narrow one (TAB. 24 fig. C2).

4. **Sesamum radiatum** Schumacher & Thonn. in Schumacher, Beskr. Guin. Pl.: 282 (1827).—Stapf in F.T.A. 4, 2: 557 (1906).—Heine in F.W.T.A., ed 2, 2: 391 (1963).—Hakki in Fl. Anal. Togo: 384 (1984). TAB. 23 fig. D & TAB. 25 fig. D. Type from the Gold Coast.
  Sesamopteris radiata (Schumacher & Thonn.) DC., Prodr. 9: 251 (1845). Type as above.
  Sesamum mombanzense De Wild. & Dur. in Pl. Thonn. Congol. 36 (1900). Type from Zaire.
  Sesamum thonneri De Wild. & Dur. op. cit.: 7 (1900). Type from Zaire.
  Sesamum talbotii Wernham in Cat. Talbot's Nigerian Pl. 73 (1913).—Hutch. & Dalz., F.W.T.A., ed. 1, 2: 244 (1931). Type from Nigeria.
  Sesamum caillei A. Chev., Expl. Bot. Afr. Occ. Fr. 1: 488 (1920), nomen.—Hutch. & Dalz. loc. cit. Type from Guinea.
  Sesamum biapiculatum De Wild. in Bull. Jard. Bot. Brux. 5: 58 (1915). Type from Zaire.

Erect annual herb, up to 1 m. high; stem simple or branched, more or less pubescent, malodorous. Leaves heteromorphic; the lower petiolate, (petioles up to 2.5 cm. long); lamina lanceolate to ovate, coarsely serrate, up to 6 × 3.5 cm.; superior leaves shortly but distinctly petiolate, lamina lanceolate or cultrate, usually entire, up to 10 cm. long; all leaves sparingly and persistently hairy and mealy glandular below. Corolla c. 3.5−4 cm. long, obliquely campanulate, purple or purplish. Capsules narrowly oblong in lateral view, c. 2.5 × 0.8 cm., pubescent, with a very short broad beak (TAB. 23 fig. 1), often with two lateral short protruberances. Seeds c. 2.5 × 1.7 mm., with a double fringe in the lower part and with radial sculptures on both faces (TAB. 24 fig. 1), black or brown when mature.

Although *S. radiatum* has not been recorded from the Flora Zambesiaca area, it is very probable that it occurs in this area, since it is widely cultivated in tropical and subtropical zones all over Africa as a crop plant for the oil which is extracted from the seeds. In habit *S. radiatum* resembles very much *S. indicum*, but it can easily be distinguished by the structure of the testa (TAB. 25 fig. D1).

Sect. **Sesamopteris** Endl., Gen. Pl.: 709 (1839).

Erect annual herbs. At least inferior leaves digitate. Inferior lip of flower not or little prominent; extra-floral nectaries single or sometimes in groups of 2−3, with minute bracts. Pollen grains usually with 6 colpi (except *S. alatum* which has 7−8). Capsules usually with prominent ribs or protruding loculi. Seeds with three wings, the two basal (lateral) ones arising from the lower fringe and bending downwards, the apical one arising from the upper fringe and bending upwards; testa faveolate, light to dark brown when mature.

Corolla 2−3 cm. long; capsule distinctly obconical in lateral view (TAB. 23 fig. E); seed wings large, up to 2 mm. long (TAB. 25 fig. E1) - - - - 5. *alatum*
Corolla 3−5 cm. long; capsule not obconical in lateral view, abaxially gibbous at the base (TAB. 23 fig. F); seeds with short (sometimes minute) wings (TAB. 25 fig F) - - - - - - - - - - - - 6. *triphyllum*

5. **Sesamum alatum** Thonn. in Schumacher, Beskr. Guin Pl.: 284 (1827).—Stapf in F.T.A. 4, 2: 559 (1906).—Bruce in F.T.E.A., Pedaliaceae: 17 (1953).—Heine in F.W.T.A., ed. 2, 2: 389 (1963).—Merxm. in Mitt. Bot. Staatss. München 3: 6.—Merxm. & Schreiber in Merxm., Prodr. Fl. SW. Afr. 131: 11 (1968). TAB. 23 fig. E & TAB. 25 fig. E. Type from the Gold Coast.
  Sesamopteris alata (Thonn.) DC. Prodr. 9: 251 (1845). Type as above.
  Volkameria alata (Thonn.) Kuntze, Rev. Gen. Pl.: 482 (1891). Type as above.
  Sesamum pterospermum R. Br. in Salt, Voy. Abyss. App. 4: 64 (1814), nomen.— Stapf in F.T.A. 4, 2: 560 (1906). Type from Ethiopia.
  Sesamum gracile Endl. in Linnaea 7: 10 (1832).—Stapf in F.T.A. 4, 2: 559 (1906). Type from Senegambia.
  Sesamum rostratum Hochst. in Flora 24 Intell.: 43 (1841), nomen.—Stapf loc. cit. Type from Sudan.
  Simsimum rostratum Bernh. in Linnaea 16: 39, 42 (1842). Type probably as above.
  Sesamum sabulosum A. Chev., Etud. Fl. Afr. Centr. Franc. 1: 229 (1913), nomen.— Bruce in F.T.E.A., Pedaliaceae: 17 (1853). Type from Zaire.
  Sesamum ekambaramii Naidu in Journ. Bombay Nat. Hist. Soc. 52: 698 (1953). Type from India.

Erect annual herb, 50−150 cm. high; stem simple or sparsely branched; glabrous except for the mucilage glands. Leaves heteromorphic; inferior leaves long-petiolate (petiole 2−7 cm. long), lamina 3−5-foliate or-partite, lobes lanceolate to narrowly linear-lanceolate, central lobe longest, 1.5−8 × 0.2−2 cm., margins entire, often undulate; superior leaves simple, lamina linear to narrowly lanceolate, 3−10 cm. long, tapering into a 1−2 cm. long petiole. Calyx deciduous. Corolla 2−3 cm. long, pink or purple, sometimes spotted red within. Capsule narrowly obconical in lateral view, 2.4−5.2 cm. long, 5−7 mm. broad in the upper part, gradually narrowed towards the base, abruptly acuminate and rostrate at the apex; beak 4−12 mm. long, straight; loculi without prominent ribs (TAB. 23 fig E). Seeds (excluding the wings) c. 2.6 × 1.5 mm., testa faveolate, with a subcircular 2−3 mm. long wing at the apex and two shorter subcircular wings at the base (TAB. 25 fig. E1).

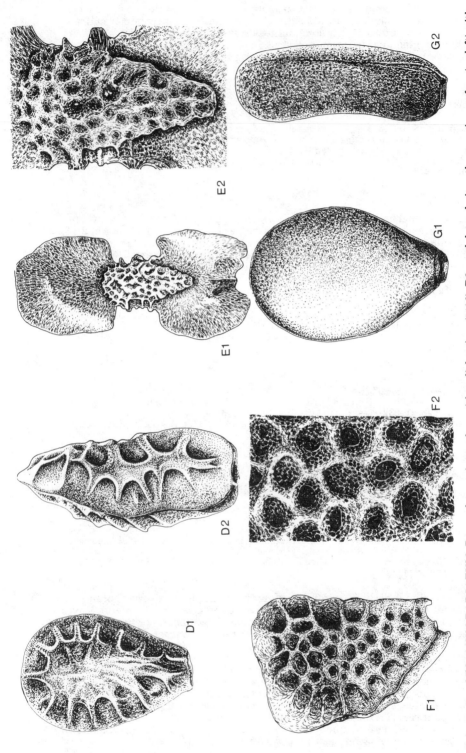

Tab. 25. D. – SESAMUM RADIATUM. D1, seed, superior surface with radial stripes (×16); D2, seed, lateral view, the superior surface (on left) with a continuous fringe (×22), D1–2 from *Flamigni* 10609. E. – SESAMUM ALATUM. E1, seed, superior surface, on both ends conspicuous wings

Caprivi Strip: Ngoma, fr. 9.i.1959, *Killick & Leistner* 3319 (M; PRE; WIND). **Botswana.** SE: 14 km. SW. of Letlhakang on Rd. to Maboane, fl. & fr. 16.ii.1960, *de Winter* 7320 (MPRE; SRGH). **Zambia.** B: near Senanga, fr. 2.viii.1952, *Codd* 7347 (PRE; SRGH). **Zimbabwe.** C: Marondera, fl. & fr. 20.iv.1941, *Dehn* 185 (M). W: Nkai Distr., Gwampa Forest Reserve, fr. viii.1955, *Goldsmith* 163/55 (BR). **Mozambique.** MS: Sefare fl. & fr. vii.1959, *Leach* 9192 (SRGH; LISC). M: Maputo, fl. & fr. 25.ii.1945, *Sousa* s.n. (LISC). Also in Namibia, S. Africa, Kenya, Ethiopia, and in W. Africa from Senegal to Chad and Sudan. Usually in sandy soils, in river beds, in grassland and in open bush.

6. **Sesamum triphyllum** Welw. ex Aschers. in Verh. Bot. Ver. Prov. Brand. **30**: 185 (1888).— Merxm. in Mitt. Bot. Staatss. München **3**: 11 (1959).—Merxm. & Schreiber in Merxm., Prodr. Fl. SW. Afr. **131** (1968). TAB. 23 fig. F & TAB. 25 fig. F. Type from Angola.
   Sesamum gibbosum Bremek. & Oberm. in Ann. Transv. Mus. **16**: 434 (1935).
   Lectotype: Botswana, Gemsbok, *van Son* T.M. 28963 (M, isotype; PRE, holotype; SRGH, isotype).

Erect annual herb, 0.25−2 m. high, bushy or simple-stemmed; glabrous except for the mucilage glands. Leaves heteromorphic; the inferior 3-foliate or 3-partite, sometimes with 1−2 additional irregular lateral segments, central lobe longest, 2−10 cm. long, with the leaflets or segments up to 2 cm. broad; superior leaves simple. Extra-floral nectaries (in the Flora Zambesiaca area) usually single. Calyx deciduous. Corolla 2.5−4 cm. long, more or less funnel-shaped, mauve. Capsules oblong in lateral view, abaxially gibbous at the base, c. 40 × 5−7 mm.; beak 6−8 mm. long, distinctly bent outwards (TAB. 23 fig. F); loculi sometimes with inconspicuous ribs. Seeds more or less horizontal in the loculi, about 2.5 mm. × 1.7 mm.; wings, especially the two basal ones, inconspicuous; testa faveolate (TAB. 25. fig. F2).

Botswana. N: Ngamiland, fr. xii.1930, *Curson* 325 (M). SW: 27 km. N. of Kang, fr. 18.ii.1960, *Wild* 5032 (SRGH; PRE). **Zimbabwe.** W: Bulawayo Distr., fr. ii.1939, *Hopkins* s.n. (SRGH). S: Beitbridge, fl. & fr. 10.i.1961, *Leach* 10666 (SRGH). **Mozambique.** M: Bay of Maputo, maize field, fr. vii.1884, *Wilms* 1070a (M). Also in Namibia, S. Angola and S. Africa. Frequent in grassland, by roadsides, growing in limestone gravel and calcareous soil.

A very variable species. Typical specimens have extra-floral nectaries occurring in groups of 3−5, large flowers (up to 5 cm. long), capsules with prominent ribs on the loculi, and seeds with conspicuous wings at both ends. In the Flora Zambesiaca area *S. triphyllum* is very similar to *S. alatum*: extra-floral nectaries usually occurring singly, flowers smaller, and capsules without prominent ribs. For correct determination it is necessary to study the seeds.

Sect. **Sesamum**

Erect annual herb. Leaves very variable; inferior leaves entire or more or less lobed to digitate, opposite or alternate. Flowers white or pinkish mauve; inferior lip not prominent; extra-floral nectaries single with conspicuous bracts. Pollengrains with 8−10 colpi. Seeds not winged but with an inconspicuous double fringe (TAB. 25 fig. 6); testa smooth, rarely slightly venose, black, brown or white when mature
· · · · · · · · · · · · · · · **7. indicum**

7. **Sesamum indicum** L., Sp. Pl. **2**: 634 (1753).—Endl., Iconogr. t. 70 (1838).—Stapf in F.T.A. **4**, 2: 558 (1906).—Bruce in F.T.E.A., Pedaliaceae: 17 (1953).—Heine in F.W.T.A., ed. 2, **2**: 391 (1963).—Ihlenf. & Seidenst. in Proc. IX Plen. Meet. A.E.T.F.A.T. Las Palmas: 53 (1979).—Hakki in Fl. Anal. Togo: 384 (1984). TAB. 23 fig. G & TAB. 25 fig. G. Type from India.
   Sesamum orientale L., loc. cit.—DC. Prodr. **9**: 250 (1845). Types from India and Sri Lanka.
   Volkameria orientalis (L.) Kuntze, Rev. Gen. Pl. **2**: 481 (1891).—Stapf in F.T.A. **4**, 2: 559. Types as above.
   Anthadenia sesamoides Ch. Lem. in Fl. Ser. **2**: pl. 6 (1846).—Stapf loc. cit. Type from Africa.
   Sesamum indicum var. *integerrimum* Engl. in Bot. Jahrb. **32**: 115 (1903). Types from Ethiopia and Zaire.
   Sesamum hopkinsii Suesseng. in Trans. Rhod. Scient. Ass. **43**: 47 (1951). Type: Zimbabwe, Marondera, 22.iii.1943, *Dehn* 755 (M).

Erect annual herb, 10−120 cm. high; stem simple or branched, obtusely quadrangular, finely pubescent to glabrescent, more or less glandular. Leaves very variable in shape, usually heteromorphic, opposite or alternate; the inferior long-petiolate (petiole 3−11 cm. long), lanceolate to ovate, 3-lobed, 3-partite or 3-foliate, 4−20 × 2−10 cm., cuneate or obtuse at the base, acute at the apex, margins often serrate; superior leaves more shortly petiolate (petiole 3−5 cm. long), narrowed, oblong-lanceolate to linear-lanceolate, 0.5−2.5 cm. broad, usually entire and narrowly cuneate at the base; all leaves thinly pubescent and more or less glandular, glabrescent. Flowers white, pink or mauve-pink with darker markings. Calyx persistent. Corolla 1.5−3.3 cm. long. Capsule narrowly oblong in lateral view, 4-sulcate, rounded at the base, 1.5−3.0 × 0.6−0.7 cm.; beak broad and short; walls of the capsule often not smooth, but covered with horizontal stripes (impressions of the seeds) (TAB. 23 fig. 6). Seeds more or less horizontal, not winged but with an inconspicuous double fringe (TAB. 25 fig. 6); testa smooth, rarely slightly venose, black, brown or white when mature.

Zambia. S: Mazabuka, 1.6 km. from Zambezi & 112 km. upstream from Kariba Gorge, fl. & fr. 12.ii.1957, Scudder 29 (SRGH). Zimbabwe. N: Binga Distr., Kariyangwe, Tsetse-Entomologist camp, fr. 15.xi.1958, Phipps 1479 (SRGH; BR). E: Mutare Distr., fl. 15.iii.1955, Chase 5520 (SRGH; BM). Mozambique. N: Lichinga, fl. 1934, Torre 574 (LISC). MS: Dondo, fl. & fr. 23.iii.1960, Leach 5230 (SRGH).
Cultivated in most tropical and subtropical countries all over the world for the oil which is extracted from the seeds.

In habit S. indicum resembles very much S. radiatum, but it can easily be distinguished by the structure of the testa (TAB. 25 figs. D, 6).

6. DICEROCARYUM Bojer

Dicerocaryum Bojer in Ann. Sci. Nat. Bot. Ser. 2, 4: 268, t. 10 (1835). Pretrea Gay ex Meisn., Pl. Vas. Gen. 1: 298 and 2: 206 (1840).—Merrill in Trans. Am. Phil. Soc. 24, 2: 355 (1935).—Abels in Mem. Soc. Brot. 25: 1−358 (1975).

Herb with a persistent woody main root, from which 5−7 prostrate annual stems bearing numerous lateral branchlets arise; whole plant subglabrous to villous and covered with mucilage-glands. Leaves very variable in shape, indument and indentation, long-petiolate to subsessile; lamina narrowly to broadly ovate, pinnatipartite to pinnatilobed or deeply crenate to serrate, the inferior surface very densely covered with mucilage-glands. Flowers solitary in the leaf axils, pendent, on long erect pedicels carrying the flowers above the plane of the leaves; flowers white to pink or mauve, occasionally violet or yellow. Corolla tube obliquely campanulate, curved, slightly saccate at the base in the adaxial side; limb sub-bilabiate; the upper and lateral lobes subequal, the anterior lobe enlarged and with dark stripes or lines of dots running down into the tube. Stamens 4, inserted near the base of the tube; staminodium absent; anthers of the shorter pair of stamens just included in the corolla tube, anthers of the longer pair of stamens just exserted from the corolla tube; thecae parallel. Ovary bilocular, false septa (not in the Flora Zambesiaca area) present or absent. Fruit woody, indehiscent, longitudinally much compressed, therefore disk-like; the central portion distinctly or only indistinctly raised, with two erect conical spines from near the centre, and the fruit easily detachable (forming a "trample bur"). Seeds dark brown to black, obovate to oblanceolate in outline, 2, (not in the Flora Zambesiaca area) 3, (not in the Flora Zambesiaca area) 4, or 5 seeds in each loculus.

A genus of 3 species, all confined to Africa.

Fruit broadly elliptic in outline, the central portion distinctly raised (TAB. 26 fig. A6); each loculus of the fruit containing 5 seeds with short funicles; leaf lamina usually narrowly ovate to ovate, pinnatipartite to pinnatilobed (Mozambique, Zimbabwe E and Malawi) or pinnatifid to deeply crenate or deeply serrate Zimbabwe, N, W, C, S); whole plant subglabrous or pubescent, the mature fruit usually glabrous
- - - - - - - - - - - - - 1. senecioides
Fruit circular to rotund in outline, the central portion only slightly raised (TAB. 26 fig. 13); each loculus of the fruit containing 2 seeds with long funicles; leaf lamina usually broadly to very broadly ovate, crenate or serrate; whole plant, including the mature fruits, densely pubescent or villous, especially in the leaf axils of the main branches - - - - - - - - - - - - 2. eriocarpum

The two species of the Flora Zambesiaca area are allopatric, but where the distribution areas are in contact (Botswana SE, Zimbabwe W), transitional forms occur which are difficult to determine if mature fruits are absent. From the leaves of all *Dicerocaryum* species a mucous infusion can be prepared which is used as a remedy in gastric and intestinal disorders and in obstetrics in cattle, and as a substitute for soap.

1. **Dicerocaryum senecioides** (Kotzsch) Abels in Mem. Soc. Brot. 25: 218 (1975). TAB. 26 fig. A. Type: Mozambique, Boror, *Peters* s.n. (B, holotype†; CGE, lectotype).
   Pretrea senecioides Klotzsch in Peters, Reise Mossamb., Bot. 1: 189−190, t. 32 (1862). Type as above.
   Pretrea artemisiaefolia Klotzsch in Peters, Reise Mossamb., Bot. 1: 189, t. 31 (1862), as "*artemisiaefolia*". Type: Mozambique, Sena, Peninsula Cabaceira, *Peters* s.n. (B, holotype†).
   Pretrea loasifolia Klotzsch in Peters, Reise Mossamb. Bot. 1: 188−189 (1862), as "*loasifolia*".−Abels in Mém. Soc. Brot. 25: 218 (1975). Type: Mozambique, Inhambane, *Peters* s.n. (B, holotype†).
   Pretrea forbesii Decne. in Ann. Sci. Nat. Bot. Ser. 5, 3: 334 (1865).−Abels in Mem. Soc. Brot. 25: 219 (1975). Type: Mozambique, "ad sinum Delagoa Africae australis", *Forbes* s.n. (K, isotype; P, holotype).

Stapf (in F.C. 4, 2: 463 (1904) and in F.T.A. 4, 2: 565 (1906)) considered *Pretrea senecioides, P. artemisiaefolia, P. loasifolia* and *P. forbesii* as conspecific with *Dicerocaryum zanguebarium* (Lour.) Merr. from Tanzania and Madagascar. All collections from the Flora Zambesiaca area belong to Subsp. *senecioides* (Abels in Mem. Soc. Brot. 25: 219 (1975)).

Stems subglabrous or pubescent, the mature fruit usually glabrous. Leaf lamina usually narrowly ovate to ovate, up to 2.7 cm. long, pinnatipartite to pinnatilobed (Mozambique, Zimbabwe E and Malawi) or pinnatifid to deeply crenate or deeply serrate (Zimbabwe N, W, C, S); petioles 3−12 mm. long. Flowers 17−27 mm. long; pedicels 15−50 mm. long. Fruit broadly elliptic in outline, 16−27 mm. long, 10−19 mm. broad, the central portion distinctly raised and 6−9 mm. high (spines excl.) each loculus containing usually 5 seeds, the central one erect, the lateral ones ascending from a very short basal placenta; funicles short. Seeds obovate in outline, c. 4.5 mm. long and 3 mm. broad.

Botswana. SE: Palapye Distr., Moeng-College, fl. 29.xi.1957, *de Beer* 510 (SRGH). **Zimbabwe.** N: Vicinity of Umvukwe Mts., fl. & fr. 29−30.iv.1948, *Rodin* 4479 (P; SRGH). W: Bulawayo, fr. v.1898, *Rand* s.n. (BM; BR). C: Marondera, fl. & fr. 12.iii.1942, *Dehn* 186 (M; SRGH). E: Chipinge Distr., on Sabi R. bank, fl. & fr. 22.iii.1959, *Phelps* 303 (SRGH). S: Zvishavane-Nyanda, fl. 15.iii.1958, *Leach* 8234 (SRGH). **Malawi.** S: Dzanjo Village, fl. & fr. 16.xi.1955, *Jackson* 1760 (SRGH). **Mozambique.** N: Between Mametil and Iuluti, fl. & fr. 12.vii.1948, *Pedro & Pedrogão* 4455 (EA). Z: Maganja da Costa, fl. & fr. 13.ix.1944, *Mendonça* 2051 (LISC). MS: Chimoio, near R. Vanduzi, fl. & fr. 28.iv.1948, *Andrada* 1200 (LISC). GI: Massengena, fl. & fr. vii.1932, *Smuts* P 363 (PRE). M: Inhaca Isl., Delagoa Bay, fl. & fr. 31.viii.1959, *Watmough* 319 (SRGH).
Also in S. Africa. In grassland or open savanna, on sand, often in disturbed vegetation and in abandoned fields, also on maritime sand dunes.

2. **Dicerocaryum eriocarpum** (Decne.) Abels in Mem. Soc. Brot. 25: 219 (1975). TAB. 26 fig. B. Type from S. Africa.
   Pretrea eriocarpa Decne. in Duchtr., Rev. Bot. 1: 517 (1846). Type as above
   Dicerocaryum eriocarpum Subsp. *eriocarpum* (Decne.) Ihlenf. in Mitt. Bot. Staatss. München 6: 599−600 (1967).−Merxm. & Schreiber in Merxm., Prodr. Fl. SW. Afr. 131: 4 (1968).−Abels loc. cit: 219. Type as above.

Stapf (in F.C. 4, 2: 463 (1904), F.T.A. 4, 2: 565 (1906)) considered this taxon as conspecific with *Dicerocaryum zanguebaruim* (Lour.) Merr. from Tanzania and Madagascar.

Stems densely pubescent or villous, especially in the leaf axils of the main branches; the mature fruits densely pubescent. Leaf lamina usually broadly to very broadly ovate, up to 22 mm. long, crenate or serrate; petioles 3−16 mm. long. Flowers 20−24 mm. long; pedicels 14−25 mm. long. Fruit circular to rotund in outline, 12−22 mm. long and 11−19 mm. broad, the central portion only slightly raised and 4−5 mm. high (spines excl.); each loculus containing usually 2 seeds which are nearly horizontally orientated and connected to the short basal placenta by long funicles; hilum of the seeds in the outer angles of the loculi. Seeds oblanceolate in outline, c. 5.5 mm. long and 2.2 mm. broad.

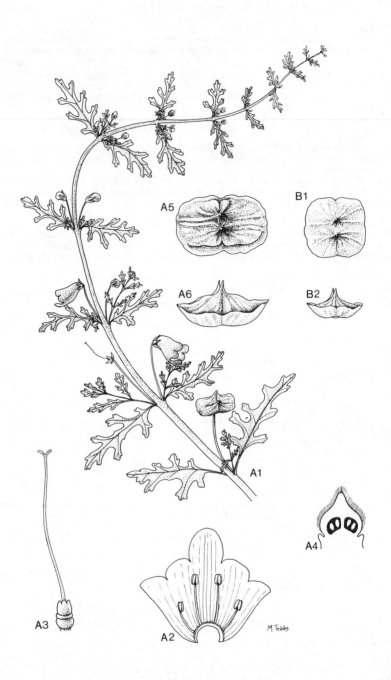

Tab. 26. A.—DICEROCARYUM SENECIOIDES. A1, habit (×½); A2, flower opened up
(×½); A3, pistil and disk (×21/2); A4, ovary in longitudinal section (×7), A1−4 after
Klotzsch in Peters, Reise Mossamb. Bot. 1: t. 32 (1863). A5, seed from above (×1);
A6, fruit, lateral view (×1); A5−6 from *Ihlenfeldt* 2129. B.—DICEROCARYUM
ERIOCARPUM. B1, fruit seen from above (×1); B2, fruit, lateral view (×1), B1−2
from *de Winter & Wiss* 4389.

Caprivi Strip: c. 16 km. from Katima Mulilo on road to Singalamwe, fl. & fr. 30.xii.1958, *Killick & Leistner* 3187 (M; WIND). **Botswana.** N: Leschuma Valley, fr. vii.1949, *Miller* B/897 (PRE). SW: Tshabong, fl. 22.ii.1960, *de Winter* 7483 (M; SRGH; W). SE: Lephepe, fr. 13.i.1958, *de Beer* 545 (SRGH). **Zambia.** B: Mongu airport, fl. & fr. 26.xi.1964, *Verboom* 1084 (HBG; S). S: Namwala Distr., fl. & fr. 1935, *Read* 22 (BM; BR; K). **Zimbabwe.** W: Hwange National Park, Main Camp, fl. 14.xi.1968, *Rushworth* 1267 (HBG; SRGH). Also in S. Africa, Namibia and Zaire. In grassland on sand, on dune slopes and river banks.

7. PEDALIUM Royen ex L.

Pedalium Royen ex L., Syst. Nat. ed. **10**: 1123 (1759).

Erect or ascending annual herb, stems simple or branched. Leaves petiolate, oblong to obovate, entire or coarsely dentate, subsucculent. Flowers solitary in the leaf axils. Corolla tube subcylindrical or narrowly funnel-shaped; limb spreading, subequally 5-lobed. Stamens included; thecae divergent; staminodium present or absent. Ovary bilocular, undivided, each loculus with 2 pendulous ovules. Fruit indehiscent, subpyramidal, 4-angled, rounded to acute at the apex and with a spreading spine at each basal angle, then abruptly contracted into a narrow collar below (TAB. 27 fig. 10); mesocarp spongy. Seeds 1—2 in each loculus.

A monotypic genus.

Pedalium murex Royen ex L., Syst. Nat. ed. **10**: 1123 (1759).—Stapf in F.T.A. 4, 2: 540 (1906).—Bruce in F.T.E.A., Pedaliaceae: 6 (1961).—Heine in F.W.T.A., ed. 2, 2: 389 (1963).—Humbert in Fl. Madag. 179 (Pedaliaceae): 32 (1971).—Hakki in Fl. Anal. Togo: 384 (1984). TAB. 27. Type from India.

Erect or ascending, sparsely glandular, 12—75 cm. high. Leaves petiolate; lamina oblong to obovate, sometimes elliptic, subsucculent; 1.5—5 cm. long, 0.8—3.5 cm. broad, coarsely dentate, particularly in the upper half, sometimes entire; rounded or truncate at the apex. Petiole 5—35 mm. long. Flowers yellow; corolla tube 2—2.5 cm. long; limb 1.5—2 cm. in diam., glabrescent or with a few hairs in the throat; lobes subcircular. Fruit 1—2 cm. long, 0.6—1 cm. broad (excl. the spines), rugose or tuberculate on the faces. Seeds narrowly oblong in outline, 6 mm. long, 1.5 mm. broad.

Mozambique. N: Pemba, fl. & fr. 21.iii.1960, *Gomes e Sousa* 4548 (K; PRE). GI: Inhambane, Vilanculos, fr. 31.viii.1944, *Torre* s.n. (LISC)
Also in Tropical east and west Africa, Socotra, Madagascar, Comores, India, Sri Lanka and Java. A saline soil indicator, sometimes a weed. The leaves are eaten as a vegetable.

8. HARPAGOPHYTUM DC. ex Meisn.

Harpagophytum DC. ex Meisn., Pl. Vasc. Gen. **1**: 298 and 2: 206 (1840). Ihlenf. & Hartm. in Mitt. Staatsinst. Allg. Bot. Hamburg **13**: 15—69 (1970). *Uncaria* Burch., Trav. Int. S. Afr. **1**: 536 (1822) nom. illeg. non Schreber, Gen. Pl. **1**: 125 (1789).

Perennial herb with several prostrate annual stems from a succulent tuberous taproot, additional tubers often present on lateral roots; whole plant glabrous or subglabrous and covered with mucilage-glands. Leaves opposite, subsucculent, petiolate, very variable in shape and often polymorphic. Flowers solitary, shortly pedicelled, erect; corolla tube nearly cylindrical, constricted at the base; limb spreading; lobes of limb subequal, circular to oblate; corolla tube light purple, pink, yellow or white outside, usually more or less yellow inside; limb purple or mauve, occasionally yellow. Stamens 4, included, inserted near the base of the corolla tube; thecae divergent; staminodium present, with apical gland. Ovary bilocular; false septa absent. Fruit a woody capsule, tardily loculicidally imperfectly dehiscent, laterally compressed, with two obtuse protuberances on each face, armed with two rows of curved arms along the edges each bearing recurved spines or the edges with two rigid wings bearing recurved spines; mature fruit easily detachable (forming a "trample burr"). Seeds

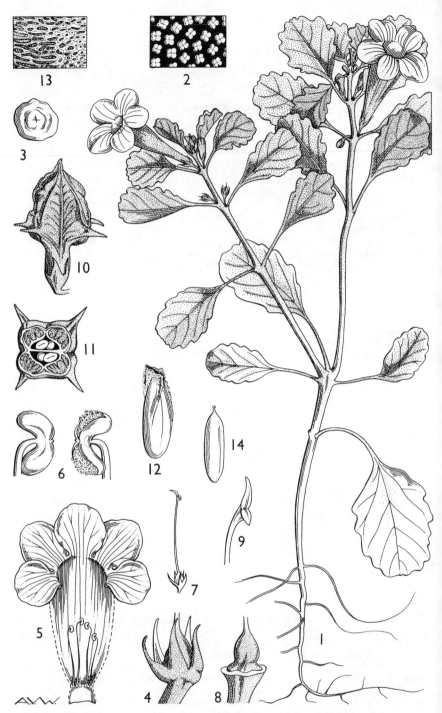

Tab. 27. PEDALIUM MUREX. 1, habit (×1); 2, portion of inferior leaf surface with mucilage-glands (×40); 3, extrafloral nectary from base of pedicel (×24); 4, calyx (×6); 5, corolla opened up (×2); 6, anthers (×16); 7, pistil and calyx (×2); 8, ovary and disk (×6); 9, stigma (×6); 10, fruit, lateral view (×2); 11, transverse section of fruit (×2); 12, seed (×4); 13, portion of testa (×20); 14, seed with the testa removed (×4), from F.T.E.A.

numerous, spreading, in 2 or 4 rows in each loculus, narrowly obovate in outline; testa irregularly reticulate, black.

A genus of 2 species, both native of southern Africa.

All taxa of *Harpagophytum* are allopatric, but where the distribution areas are in contact (especially in Botswana N, SE, and Zimbabwe S), numerous transitional forms occur. Without mature fruits, correct identification may be difficult, since due to the frequent polymorphy of leaves, leaf characters given in the keys below may be unreliable; attention to the distribution of the taxa may be helpful in the identification of incomplete material.

Fruit with 4 rows of curved arms bearing recurved spines, the length of the longest arm exceeding the width of the capsule proper by up to five times; seeds 25−30 in each loculus, in 4 rows; leaves usually with a petiole about two thirds of the length of the lamina; leaves usually narrowly ovate to ovate - - - 1. *procumbens*

Fruit with 4 rows of slightly curved arms bearing recurved spines, the length of the longest arm not exceeding the width of the capsule proper, or fruit with 4 rigid wings bearing recurved spines, the wings sometimes dissected into short arms in the upper portion of the fruit; seeds 10−15 in each loculus, in 2 rows; leaves usually with a petiole less than half the length of the lamina; leaves usually broadly ovate to ovate - - - - - - - - - - - - - - 2. *zeyheri*

1. **Harpagophytum procumbens** DC. ex Meisn., Pl. Vasc. Gen. 1: 298 and 2: 206 (1840). Type from S. Africa.
   Uncaria procumbens Burch. Trav. Int. S. Afr. 1: 536, fig. p. 529 (1822), nom. illeg.—DC., Prodr. 9: 257 (1845).—Decne. in Delessert, Ic. sel. pl. 5: 39−40, tab. 94 (1846).—Stapf in F.C. 4, 2: 458 (1904).—Stapf in F.T.A. 4, 2: 548 (1906).—Ihlenf. in Mitt. Bot. Staatss. München 6: 600−602 (1967).—Merxm. & Schreiber in Merxm., Prodr. Fl. SW. Afr. 132: 4−5. Type from S. Africa.
   Harpagophytum burchellii Decne. in Duchtr., Rev. Bot. 1: 516 (1846), nec Decne. in Ann. Sci. Nat. Bot. Ser. 5, 3: 329, tab. 11 fig. 7 (1865).—Stapf in F.C. 4, 2: 458 (1904). Type from S. Africa.

Perennial herb with several prostrate annual stems from a succulent taproot, with additional tubers on lateral roots. Leaves narrowly ovate to ovate, up to 65 mm. long and 40 mm. broad; petiole 30−45 mm. long; lamina usually pinnatilobed, with 3 or 5 main lobes (terminal lobe included), sometimes polymorphic. Limb of corolla purple or (often in subsp. *transvaalense*) yellow, 25−40 mm. in diam.; tube usually light purple or pink, occasionally white, outside, usually more or less yellow inside, 50−60 mm. long. Fruit with 4 rows of curved arms bearing recurved spines, the length of the longest arm exceeding the width of the capsule proper, the total diam. of fruit up to 15 cm. Seeds 25−30 in each loculi, in 4 rows.

Leaves usually narrowly ovate, pinnatilobed with 5 main lobes; fruit usually with 4 arms in each row (sometimes only three, but then lowest arm of each row much broader and often imperfectly divided), the length of longest arm usually two to five times the width of the capsule proper; Botswana SW. and adjacent parts of Botswana N. and SE. - - - - - - - - - - subsp. *procumbens*
Leaves usually ovate, pinnatilobed with 3 main lobes; fruit usually with 3 arms in each row, the length of the longest arm usually not exceeding twice the width of the capsule proper; Zimbabwe S. - - - - - subsp. *transvaalense*

Subsp. **procumbens** TAB. 28 fig. C1−3

Botswana. N: Ngamiland, c. 24 km. from Matsibe Tsetse Camp, Okavango, fr. 17.iii.1961, *Richards* 14760 (K). SW: 25 km. NW. Hukuntsi, fl. & fr. ii.1979, *Skarpe* 319 (K; PRE). SE: N. of Lephephe, fl. & fr. ii.1982 *Snyman & Noailles* 217 (PRE).
   Also in Namibia and S. Africa. In deep red or brown sand in bush savanna, especially in overgrazed places.

Subsp. **transvaalense** Ihlenf. & Hartm. in Mitt. Staatssinst. Allg. Bot. Hamburg 13: 57 (1970). TAB. 28 fig. D1−4. Type from S. Africa.

Zimbabwe. S: Beitbridge, near Limpopo R., fl. & fr. 15.ii.1955, *Exell, Mendonça & Wild* 429 (BM; LISC; SRGH).
   Also in S. Africa. In deep sand in open woodland.

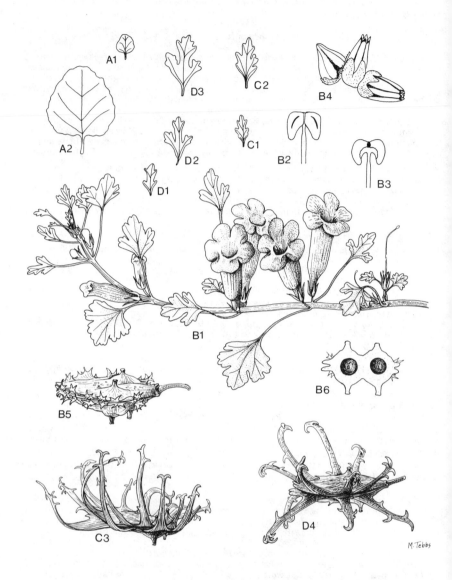

Tab. 28. A.—HARPAGOPHYTUM ZEYHERI Subsp. SUBLOBATUM. A1, leaf at
beginning of season on annual shoot (×⅓); A2, leaf at mid-season (×⅓), A1−2 after
Ihlenf. & Hartm. loc. cit. (1970); B.—HARPAGOPHYTUM ZEYHERI Subsp.
ZEYHERZI. B1, habit (×⅓), based on photograph (*Ihlenfeldt* 2528) (length of petioles
not typical, usually much shorter); B2, anther adaxially (×10); B3, anther abaxially
(×10); B4, group of three extrafloral nectaries from base of pedicel (×8), B1−4 after
Ihlenf. & Hartm. loc. cit.; B5, fruit, lateral view (×⅓); B6, fruit, transverse section
(×⅓), B5−6 from *Ihlenfeldt* 2549; C.—HARPAGOPHYTUM PROCUMBENS Subsp.
PROCUMBENS. C1, leaf at beginning of season (×⅓); C2, leaf from mid-season (×⅓),
C1−2 after *Ihlenf. & Hartm.* loc. cit.; C3, fruit, lateral view (×⅓) from *Ihlenf. &
Hartmann* 4004; D.—HARPAGOPHYTUM PROCUMBENS Subsp.
TRANSVAALENSE. D1, leaf of beginning of season (×⅓); D2, leaf in mid-season
(×⅓); D3, leaf at end of season (×⅓), D1−3 from *Ihlenf. & Hartm.* loc. cit; D4, fruit,
lateral view (×⅓), from *Ihlenfeldt* 2160.

The tuber of *Harpagophytum procumbens* are collected, fermented and dried. Used as a drug it is chewed or an infusion is prepared. It is applied as a remedy in various diseases, such as fever, skin lesions, rheumatism, intestinal disorder, headache, arteriosclerosis and diseases of liver, kidney and bladder.

2. **Harpagophytum zeyheri** Decne. in Ann. Sci. Nat. Bot. Ser. 5, **3**: 329 (1865).—Stapf in F.C. **4**, 2: 458 (1906).—Stapf in F.T.A. **4**, 2: 458 (1906).—Ihlenf. in Mitt. Bot. Staatss. München **6**: 600—602 (1967).—Merxm. & Schreiber in Merxm., Prodr. Fl. SW. Afr. **132**: 5 (1968). Type from S. Africa.
Harpagophytum peglerae Stapf in Bolus, Trans. S. Afr. Phil. Soc. **16**: 398 (1906).—De Wild., Pl. Nov. Herb. Hort. Then. **2**: 20—21, tab. 64.—Ihlenf. & Hartm. in Mitt. Staatsinst. Allg. Bot. Hamburg **13**: 57 (1970). Type from S. Africa.

Perennial herb with several prostrate annual stems from a succulent taproot, which often forms a thick tuber. Leaves petiolate; lamina usually 20—40 mm. long, rarely up to 60 mm., and 20—35 mm. broad, very broadly ovate to ovate, pinnatilobed to crenate, often polymorphic; petiole 10—15 mm. long (usually less than half the length of the lamina); limb of corolla purple or yellow, 15—25 mm. in diam.; tube purple, violet, pink or yellow outside, usually more or less yellow inside, 40—50 mm. long. Fruit with 4 rows of slightly curved arms bearing recurved spines, the length of longest arm not exceeding the width of the capsule proper, or fruit with 4 rigid wings bearing recurved spines, the wings in the upper portion of the fruit sometimes dissected into short arms; capsule proper 40—50 mm. long and 18—23 mm. broad, the total length of fruit up to 105 mm. and the total width up to 70 mm. Seeds 10—15 in each loculus, in two rows.

Leaves usually ovate, very often polymorphic, pinnatilobed, pinnatifid or crenate, but at least some leaves of the plant distinctly pinnatilobed with usually 3 main lobes (terminal lobe included); fruit with rigid wings or short arms, the width of wings or the length of longest arm up to half the width of the capsule proper; 3—4 extrafloral nectaries often present at the base of each flower; Botswana SE., Mozambique M.
- - - - - - - - - - - - - - - subsp. *zeyheri*
Leaves usually very broadly ovate to ovate, not polymorphic, crenate to pinnatifid; fruit usually with broad arms, the length of the arms equaling half the width or the full width of the capsule proper; extrafloral nectaries rarely more than two at the base of each flower; Botswana N., Caprivi Strip, Zambia B., S., Zimbabwe W., S.
- - - - - - - - - - - - - - subsp. *sublobatum*

Subsp. **zeyheri** TAB. 28 fig. B1—6

Botswana. SE: Mahalapye, fl. & fr. 21.xii.1957, *de Beer* 536 (BR; K; PRE; SRGH).
Mozambique. M: Canicado 16 km. towards Lagoa Nova near Mapulanguene, fl. 19.xi.1970, *Correia* 2053 (WAG).
Also in S. Africa. In deep, sometimes loamy sand in grassland, especially in overgrazed places.

Subsp. **sublobatum** (Engl.) Ihlenf. & Hartm. in Mitt. Staatsinst. Allg. Bot. Hamburg **13**: 58 (1970). TAB. 28 fig. A1—2. Type from Angola.
Harpagophytum procumbens f. *sublobatum* Engl. in Warburg, Kunene-Samb.-Exped. Baum : 370 (1903). Type as above.
Harpagophytum procumbens var. *sublobatum* (Engl.) Stapf in F.T.A. **4**, 2: 548 (1906). Type as above.

Caprivi Strip: Lizauli, fl., 2.i.1959, *Killick & Leistner* 3257 (K; M; PRE; WIND).
Botswana. N: 20 km. S. of Hyaena camp on Linyanti R., fl. & fr. 16.x.1972, *Biegel, Pope & Russell* 4073 (K; PRE; SRGH). **Zambia.** B: 3.2 km. W. of Kalabo, fl. 13.xi.1959, *Drummond & Cookson* 6433 (SRGH). S: Livingstone, fl. s.dat., *Rogers* 7517 (Z). **Zimbabwe.** W: Hwange National Park, 2.4 km. E. of Shapi Camp, fl. & fr. 27.ii.1969, *Rushworth* 1574 (K; SRGH). S: Mwenezi Distr., Gone-Re-Zhou Game Reserve, fr. 2.vi.1971, *Ngoni* 151 (SRGH).
Also in Angola and Namibia. In Kalahari sand in open woodland.

INDEX TO BOTANICAL NAMES